Praise for *Changing Ti*

A needful and accessible book of soberly optimistic ecology as it is a condemnation of colonialist appropriation of territory and thought. By challenging Eurocentric science to pay deeper attention to traditional knowledge, Frid bridges the artificial gap between ways of human behavior on the planet with lyricism and respect.

— Anna Badkhen, author, *Fisherman's Blues* and *Walking with Abel*, and co-editor, *Changing Tides*

Seamlessly blends impeccable science with indigenous knowledge and offers a hopeful call to action to save our planet and ourselves. Beautifully written, poignant, and mind expanding, this outstanding book deserves a broad global audience so that we can begin right now to find our way back to our place in nature.

— Marc Bekoff, Ph.D. author, *Rewilding Our Hearts* and *The Animals' Agenda: Freedom, Compassion, and Coexistence in the Human Age*

This is a beautifully written book about the people, plants, animals and spirits that inhabit the British Columbia coast, a habitat under great strain from climate change and other human impacts. But this is not a doom and gloom tale; Frid marries lyrical writing, compelling stories and sharp ecological and cultural insights to provide an uplifting vision of how scientific and Indigenous ways of knowing working together could provide a way forward to prevent impending environmental collapse.

— Mark L. Winston, Professor and Senior Fellow, Simon Fraser University's Morris J. Wosk Centre for Dialogue, and author, *Bee Time: Lessons From the Hive*, winner of the 2015 Governor General's Literary Award for Nonfiction.

A positively uplifting read! *Changing Tides* offers not only a vision for a buoyant planetary future but also a carefully defended argument to believe in it. Frid's stories reveal how Indigenous knowledge and science provide a potent combination to guide us through this time of great uncertainty.

— Chris Darimont, Raincoast Chair of Applied Conservation Science, University of Victoria

The narrative here reaches far beyond the natural world. It's a story about kindness and respect, inspiration and reward. If one is interested in doing better for our collective futures, *Changing Tides* needs to be digested if for no other reason than valuable lessons from our past and present.

— Joel Berger, scientist and author, *Extreme Conservation*

How is it possible to encapsulate the natural and cultural history of a coast, concerns for the future, the joy of being with people you admire in a place you love, and the qualities of an ecosystem burgeoning with intricate relationships, all in a single volume? That's what Alejandro Frid has done, in this engaging, informative and life affirming book about his work on the central coast of British Columbia.

— Nancy Turner, CM, OBC, FRSC,
Distinguished Professor Emeritus, University of Victoria

A beautifully crafted journey into how we can change our destructive global culture—and who we can learn from. Quite simply, this is what real hope looks like.

— J. B. MacKinnon, author, *The Once and Future World*

Describing the wisdom from traditional and modern knowledge, Alejandro Frid brilliantly outlines a pathway for a viable and enduring future. Frid encourages us to change our cultural story so that we can manage the inevitable ecological changes due to the climate crisis.

— Andres R. Edwards, author, *Renewal* and *The Heart of Sustainability*

In this beautifully rendered book, *Changing Tides*, Alejandro Frid addresses how we as humans can live and act in the face and fear of climate change. This book offers hope and paths forward, possibilities both place specific and universal, deeply personal yet holding promise for humanity.

— Dr. Mehana Blaich Vaughan, author, *Kaiāulu: Gathering Tides*

CHANGING TIDES

An Ecologist's Journey to Make Peace with the Anthropocene

ALEJANDRO FRID

Cover design by Diane McIntosh.
Front cover Drummer Image: Alejandro Frid
(See note about cover image opposite.)

Fish illustration © iStock
All photos © Alejandro Frid unless otherwise noted.
All other artwork © Michael Nicoll Yahgulanaas (mny.ca)
unless otherwise noted.

Printed in Canada. First printing October 2019.

Inquiries regarding requests to reprint all or part of *Changing Tides* should be addressed to New Society Publishers at the address below.
To order directly from the publishers, please call toll-free (North America) 1-800-567-6772, or order online at www.newsociety.com

Any other inquiries can be directed by mail to

New Society Publishers
P.O. Box 189, Gabriola Island, BC V0R 1X0, Canada
(250) 247-9737

LIBRARY AND ARCHIVES CANADA CATALOGUING IN PUBLICATION

Title: Changing tides : an ecologist's journey to make peace with the anthropocene / Alejandro Frid.

Names: Frid, Alejandro, 1964– author.

Description: Includes bibliographical references and index.

Identifiers: Canadiana (print) 2019014808X | Canadiana (ebook) 20190148101 | ISBN 9780865719095 (softcover) | ISBN 9781550927023 (PDF) | ISBN 9781771422987 (EPUB)

Subjects: LCSH: Ethnoscience.

Classification: LCC GN476 .F75 2019 | DDC 306.4/2—dc23

Funded by the Government of Canada | Financé par le gouvernement du Canada | Canada

New Society Publishers' mission is to publish books that contribute in fundamental ways to building an ecologically sustainable and just society, and to do so with the least possible impact on the environment, in a manner that models this vision.

FRONT COVER IMAGE: Edward Johnson, originally from the Esk'etemc Nation and married into the Tŝilhqot'in Nation, drums during a ceremony at Teztan Biny, a lake sacred to the Tŝilhqot'in peoples of interior British Columbia. The drum was created by the great artist Eugene Hunt (1946–2002) of the Kwakwaka'wakw Nation of Vancouver Island and the adjoining mainland; my father, Samuel Frid (1935–2010), acquired it in the 1980s and passed it on to me in the early 1990s. It remained with my family until September of 2018, when I gave it to Cecil Grinder and Doreen William, both Tŝilhqot'in, to celebrate their wedding. Outside the photograph, Cecil and Doreen stand by the shores of Teztan Biny. The web of social, geographic, and cultural relationships held within this image reflects the trade economy, cross-pollination, and adaptability that are integral to First Nations. These are all major themes of this book.

For Gail and our forest time

For Twyla Bella and her stories to be

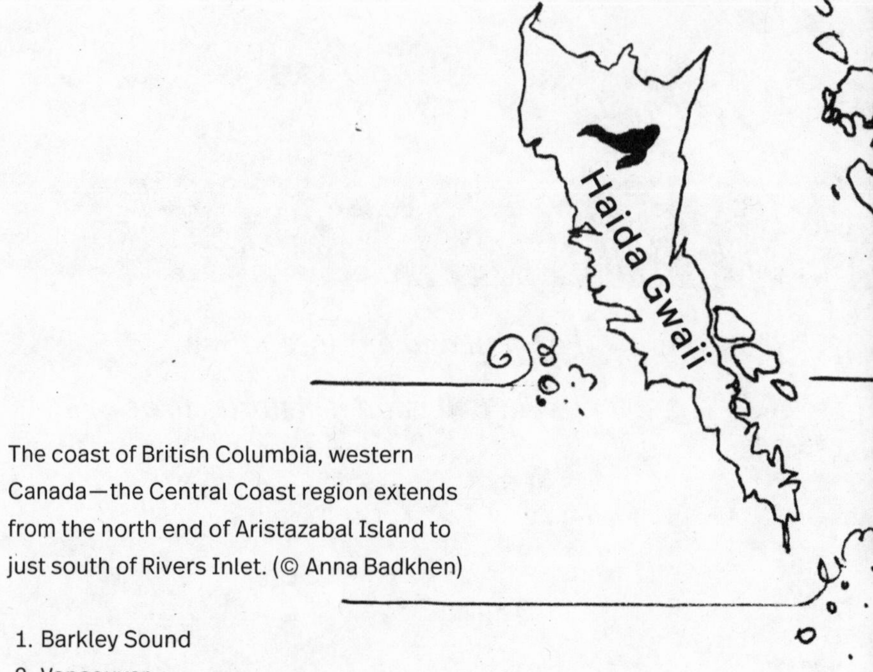

The coast of British Columbia, western Canada—the Central Coast region extends from the north end of Aristazabal Island to just south of Rivers Inlet. (© Anna Badkhen)

1. Barkley Sound
2. Vancouver
3. Bowen Island and Howe Sound
4. Teztan Biny
5. Wuikinuxv Village (Wuikinuxv Nation), Rivers Inlet, Oweekeno Lake
6. Bella Bella (Heiltsuk Nation)
7. Bella Coola (Nuxalk Nation)
8. Chugways
9. Spiller Channel (Herring walls)
10. Kitasu Bay
11. Klemtu (Kitasoo/Xai'xais Nation)
12. Aristazabal Island
13. Mussel Inlet
14. Kynoch Inlet
15. Namu

N
S
13
14
12
10
11
9
8
6
15
5
4
Vancouver Island
3
2
1
299

Contents

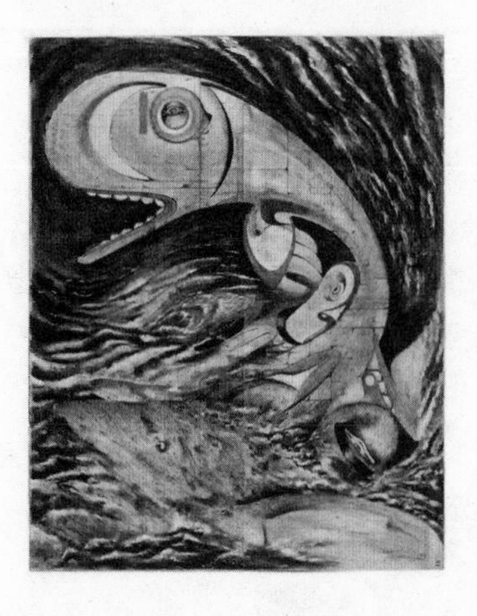

Preface

Like many of my scientific colleagues, I am often overwhelmed. Climate change, ocean acidification, species extinctions: we contemplate these difficult issues constantly. I know well what it is like to just want to give up.

It seems so easy: losing faith in humans. It promises relief from struggle and responsibility. Yet, whenever I have gone there, I have also felt empty. Claustrophobic. Horribly hollow.

And, apparently, I am too chicken to stomach those feelings. Whenever I have allowed myself to sink into cynicism, I have—invariably—jolted myself out of my catatonic state before hitting bottom and resumed swimming towards shore.

As an ecologist working on marine conservation with modern Indigenous peoples of the Northeast Pacific Ocean, I live at the crossroads of different world views and ways of knowing that, I believe, capture some of the best that humans have to offer to ourselves and to our non-human kin. We already have set in motion such rapid and ineluctable changes to our planet that both the traditional knowledge of Indigenous peoples and science will have to remain fluid and adaptive in order to not become obsolete. Both knowledge systems are designed to do exactly that. When combined synergistically, they can provide us with the tools we need to keep learning as change continues and accelerates—helping us connect with fundamental pieces of reality in ways that might allow us to remain our essential selves.

This book is my personal journey through the interface of science and traditional Indigenous knowledge. It is the story of why, despite the apparent evidence trying to talk me into doing otherwise, I believe in us.

Different cultures—collective ways of perceiving, knowing, creating, and behaving in the world—are combining today in ways that our ancestors would have welcomed. That is the challenging gift that accompanies the ongoing transformation of our planet into something that, in many ways, would be unrecognizable to those who lived before us, even in the near past.

I do not deny the losses that accompany that transformation. A planet in which wild salmon and ancient rainforests are being diminished is something to mourn. Yet I also like to think that, if they could catch a glimpse of our modern world, departed ancestors from Indigenous cultures of the northeast Pacific Ocean would recognize the continuity of many of their fundamental legacies, such as adaptability to change and the responsibilities of knowing how to give and how to receive a gift. And, above all, kinship.

These legacies, and more, are held within the works that artist Michael Nicoll Yahgulanaas gifted to this book. Michael described this gift as a symbol of the unprecedented solidarity that exists today among many Indigenous peoples and of the alliances that are being formed between First Nations and settlers who came from away. Michael is Haida. His people and the Central Coast First Nations I feature in this book were once dangerous enemies but are now fierce friends. And despite the past and ongoing crimes perpetrated by some settlers and their governments against the original inhabitants of the land, today millions of people from different Indigenous Nations and from settler groups are working together, globally, to fulfil our common obligations of respect, gratitude, and reciprocity towards all living things.

When Michael offered the gift of his art for this book, it reaffirmed for me that we live in fortunate times.

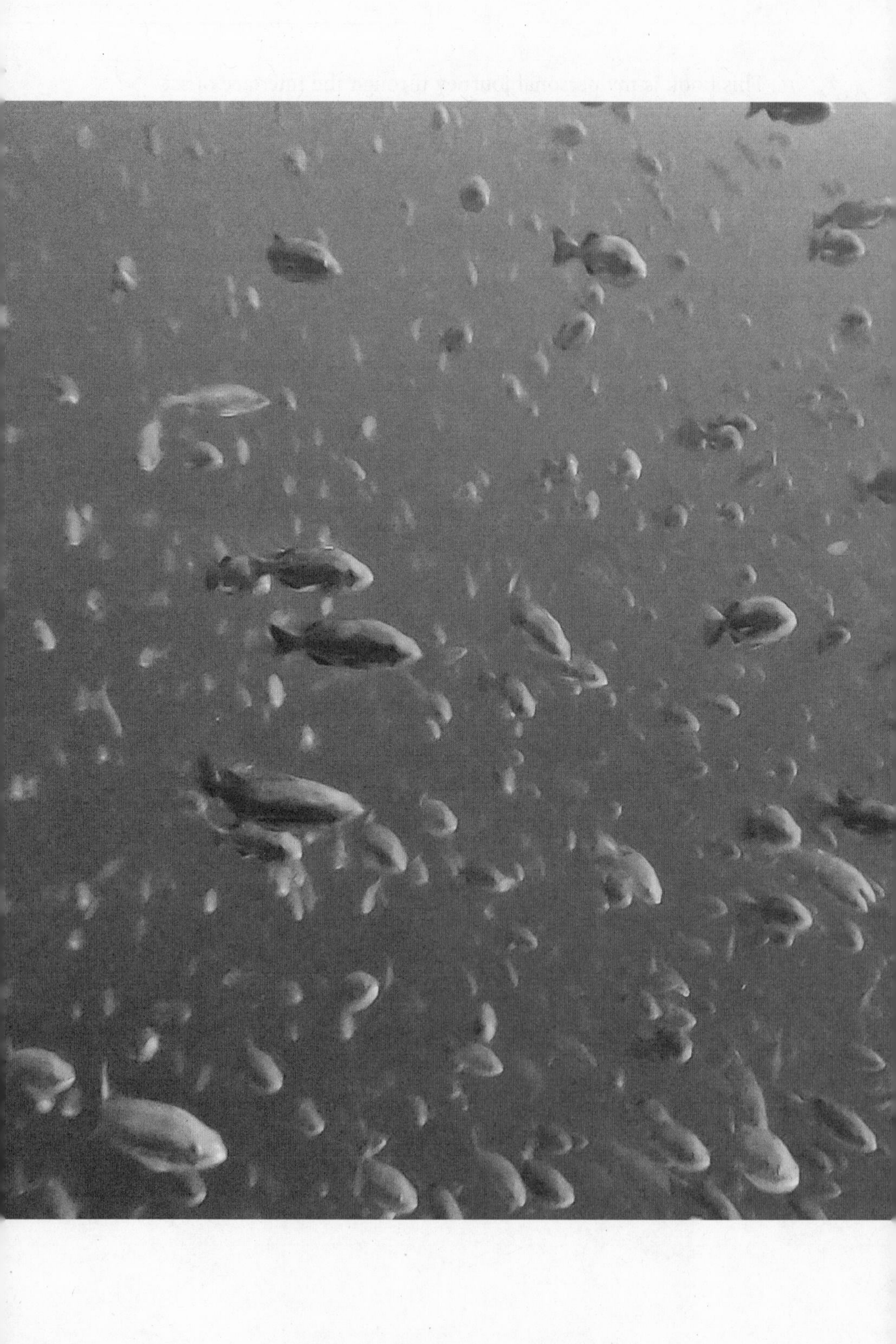

1

Gravity Suspended

A swift transition between worlds. That is how I experience the start of a research dive into the ocean—when I roll backwards from the gunwale of a small boat and, for a second, my vision traces an arc across the sky that culminates inside a burst of white bubbles. And then: the transparency, or murkiness, of the underwater world. The air trapped within my dry suit bobs me up to the surface, but only briefly. As I press the air release valve, my buoyancy steals away. And I sink…

That is the instant when all heaviness vanishes: the weight of my tank and other gear, of my body itself, and perhaps my mind too. I am now free to plummet, float in place, spin, rise—to act as if the very existence of gravity has been suspended.

And maybe just before I plunged into the water, the wind or boat engines were loud. Maybe wolf howls from the nearby forest pierced the air. Whatever the sounds were above the surface, they are now gone, replaced by the rhythm of my own breathing.

I always pause to acknowledge this transformation, this shift in perspective that allows me to access a unique freedom of the body and the mind. I soak it in so that it may stay with me throughout the busyness I am about to face when surveying rockfish—a genus of long-lived, marine fish that are easy to overexploit and culturally significant to Indigenous peoples who live along the coast of the northeast Pacific Ocean.

The surveys are part of the long-term and collaborative research I conduct with four First Nations along the Central Coast of what is now known as British Columbia, in western Canada. The ancestral territories of these Indigenous groups are made up of lush temperate rainforests and vertical granite walls that rise from a rich ocean to become corrugated mountains, where you can stand on a glaciated peak and look, almost straight down, into estuaries and rivers where the white-coated spirit bear (which is found nowhere else in the world) and other predators—grizzly bears, wolves, and wolverine—feed seasonally on large numbers of spawning salmon. These animals scatter fish carcasses among tall Sitka spruces that are nurtured by the decomposing flesh, linking the high seas that fed the salmon with forests that sustain myriad species of birds, insects, plants, lichens, and fungi. And, if standing on that peak, you raise your gaze slightly and the day happens to be clear, you will see fjords give way to islands large and small, some mountainous and heavily forested, some flat and strewn with bogs and ponds that resemble the subarctic, some little more than windswept rocks where seabirds nest and sea lions haul themselves out to rest. The islands extend far out to sea, sparkling in the sun amidst the breaking waves of the northeast Pacific.

I am the ecologist and science coordinator for the Central Coast Indigenous Resource Alliance, which the Wuikinuxv, Heiltsuk, Nuxalk, and Kitasoo/Xai'xais First Nations have created to join forces in the proactive management of resources within their territories. Our studies of rockfish and other marine organisms—including Pacific herring and Dungeness crabs—are part of an effort to support conservation and fishery management by cultures that are both grounded in ancient traditions and very much part of the modern world. While contemporary life has brought opportunities to these coastal cultures, it has also brought challenges. These include rapid climate changes that reassemble biological communities into unprecedented configurations of species; industrial fisheries that carry off vast amounts of fish very quickly; and logging, which has destroyed parts of the rainforest and clogged some nearshore ecosystems with wooden debris that smother marine life and deplete life-sustaining oxygen dissolved in seawater. Yet the notion of managing people's behavior in the face of the human potential to destroy ecosystems is not new to my Indigenous friends and colleagues. Their tradition has always recognized our destructive capacity and—more importantly—our power to preempt it.

It is through my time with First Nations—in the field, in communities, and in the city boardrooms of Vancouver—that I have come to understand that the seemingly wild coast of British Columbia has been home to very large populations of technologically sophisticated cultures for thousands of years. In the process, I've come to appreciate that, despite having had the capacity to deplete rockfish and other species, they did not.

These facts tell me something profound about humans. Something that many of us have failed to recognize, and that may be essential for global society to avert the worst of our current, and very pressing, climate change and biodiversity crises.

Rockfish are a mirror for how we use—or don't use—some of our best human qualities.

The genus *Sebastes* (as rockfishes are scientifically called) stands out for its many species in which individuals can live to be centenarians, occasionally twice over. Chief among them are rougheye rockfish: solitary shrimp- and fish-eaters known to reach the astounding age of 205 years. Rougheyes can grow to half the size of a tall person. They prefer the relative darkness of greater depths—200 to 400 meters beneath the surface—which means that I will never dive among them.

The long-lived species that I know best are yelloweye and quillback rockfish. Yelloweyes are a beautiful orange color punctuated by one or two white stripes. They have been aged to 121 years and can grow almost as large as rougheyes. Quillbacks can live nearly a century and grow to about half the length of yelloweyes. They look like miniature interstellar events—bright yellows merge into deep blacks, and white quills shoot straight up from their backs. During our dive surveys, we commonly see young and middle-aged individuals of these species. Yet the oldest and largest fishes are found much deeper than my colleagues and I can dive, so we study them through fishery catches and video from remotely operated cameras.

Quillbacks and yelloweyes have occupied much of my attention because their cultural and biological contexts converge in important ways. Both species are highly prized in the traditional diets of Coastal First Nations. They also have—like other species in which individuals live a century or more—a very slow life history that makes them very vulnerable to overfishing: long-lived individuals also grow slowly and take a long time to start reproducing, one to two decades in the case of many rockfishes. Having a slow life history also means that, once sexually mature, females reproduce annually throughout their very long lives. Critically, larger and older mothers are disproportionately more fecund than smaller, younger mothers. A 50-centimeter-long quillback rockfish will birth 800,000 larvae; that is seven times more offspring than a quillback mother that is only one third smaller. The greater fecundity of larger mothers is critical, because larvae have a tough life that is often short. Immediately upon birth, these baby fish—which are smaller than my pinky nail, have huge orange eyes and a translucent skin that makes the spine and innards visible to an observer—are thrust into the predator-laden ocean to fend for themselves. Most get eaten or starve within a few months. Odds are that very few or no larvae from a cohort will survive into reproductive adulthood. In some rockfish species, older mothers are important not only because of their hyper-fecundity but also because they give birth earlier in the year than younger females. This variation between different-aged mothers extends the length of the reproductive season and increases the chances that larvae will be born when their main food—tiny invertebrates known as zooplankton—are most abundant. The more zooplankton they can access shortly after birth, the faster the larvae will grow and be able to avoid predators, including larger rockfishes.

These life history characteristics are both fascinating and worrisome. Given their higher reproductive capacity, big, old mothers are essential for the persistence of most rockfish species, yet modern fisheries target the biggest and oldest fish, effectively shrinking the average size and age of fish populations. The consequences can be misleading: when examining the impact of a fishery we may find that lots of fish are still around, and perhaps even pat ourselves on the back for not depleting the stock. But if we take a more nuanced view and realize that those remaining fish are all young and small, we will recognize a tenuous situation called longevity overfishing—the loss of big old fish that can be a precursor to a stock collapse, or a barrier to rebuilding and maintaining a sustainable fishery.

Yet not all rockfish species are characterized by extremely long life spans, which is important for fishery sustainability. Yellowtail, widow, and black rockfishes, for instance, have maximum ages of only 50 to 60 years, which means that their life history is just plain slow (as opposed to *very* slow), and therefore these species are less vulnerable to overfishing than longer-lived species.

The diversity of rockfish life spans correlates with a diversity of lifestyles that can make the busyness of our research dives hard-hitting and immediate. My colleagues and I often have the cosmic experience of descending midwater through mixed-species schools that include many hundreds of yellowtail, widow, or black rockfishes, plus a few individuals from similar species, like dusky and deacon rockfish. These schools create a moving mass of yellow and black bodies, with a smattering of whitish, greenish, and blue tones thrown in, hovering far off the bottom in crystal clear water, sometimes preying on herring and other small fishes. Upon reaching the reef we encounter a very different set of species—including quillback and yelloweye—among boulders or inside bedrock crevices, often solitary. After settling on a reef, adults of these longer-lived species will stay near the bottom and move distances of only a few hundred meters for the rest of their lives. In other words, longer-lived rockfishes are archetypal "sitting ducks" for depletion by anybody with a boat, basic fishing gear, time, and disregard for the stewardship of resources that the ocean offers us. In a world where humans have derived much of their food from the ocean, longevity overfishing in general, and the status of rockfish populations in particular, are litmus tests for the human capacity to destroy—or choose to conserve.

Given their vulnerability to human exploitation, why are long-lived rockfishes still here if First Nations have had plenty of time and capacity to overfish them? Indigenous people have been harvesting rockfishes in British Columbia for at least 9,100 years, perhaps longer, given that humans have occupied the region for at least 14,000 years. And rockfish fisheries have always been geographically widespread. Archaeologists studying middens—sites within former villages or camps where shells, bones, and similar food remains accumulated over time—have found rockfish bones at nearly two thirds of the sites examined throughout British Columbia and adjacent areas of Oregon, Washington, and Southeast Alaska. Importantly, rockfishes were not a rare treat. Of 17 types of fish remains in middens, rockfish bones are the sixth most abundant throughout the region, and the fifth most abundant in British Columbia's Central Coast. This means that rockfishes were a predictable, year-round staple food, fished for millennia by large and dense human populations.

The middens of Barkley Sound, in southern British Columbia, are well-studied and particularly illustrative of the sustainability of Indigenous fisheries for rockfish. The sound covers 800 square kilometers, most of which is water, and its pre-colonial population is estimated to have been about 8,500 people: twice the current population in the area. By analyzing DNA from fish bones in middens, archaeologists recently identified continuous consumption of at least 12 rockfish species over the course of 2,500 years. I find this remarkable, given that many rockfishes have a slow life history and are vulnerable to overfishing.

Even more notably, prior to colonization, Indigenous fishers were not technologically limited in their fishing capacity. Granted, their potential to overexploit was lower than that of modern, fossil-fuel-powered industrial fleets, but it was high enough to inflict serious damage. Dozens of paddlers propelled massive canoes far offshore, hunting whales, Pacific bluefin tuna, and other species. Sticking closer to shore and home, where many rockfish species can be fished, would have been easy, especially because people had very strong longlines made from cured bull kelp, hooks designed to target specific types of fish, and basket traps. That technology, coupled with a large human population, could have wiped the reefs clean of all their rockfish. Yet that sort of depletion did not happen.

The ancient DNA data for Barkley Sound suggests that people associated with the middens rarely consumed the slower-growing, longer-lived species (such as yelloweye and quillback rockfish) that are easier to deplete. Instead, they ate mostly species with shorter life-spans—such as yellowtail, widow, and black rockfishes—which grow faster, begin reproducing earlier in life, and withstand fishing pressure better than longer-lived species. These findings are rather puzzling to me. In our modern times, Coastal First Nations cherish the taste of quillback and yelloweye, and formal interviews with elders from their communities reinforce the cultural and nutritional role that these species have had over their lifetimes. So why do we not see a higher proportion of longer-lived species in the ancient DNA data set?

The archaeological data from Barkley Sound suggest that fishers kept their gear above the bottom and instead fished midwater, particularly near kelp beds, where smaller fish provide ample prey for yellowtail, black, and similar rockfish species. In other words, fishers appear to have made the deliberate choice of exploiting shorter-lived rockfishes at higher rates than longer-lived rockfishes, actively mitigating potential impacts on the most vulnerable species and allowing for sustainability over the course of millennia.

True, evidence from middens cannot distinguish whether conservation of long-lived rockfish that live along the bottom was active or passive. After all, fishing for mid-water species at shallower depths requires less effort and a lower risk of snagging and damaging gear, which may have influenced the fishers' behavior. But the fact is that other species that live in deeper water and along the bottom were targeted. For instance, the remains of Pacific halibut—a delicious flatfish that can grow to two and a half meters long and weigh one third of a ton—occur in one fifth of the middens in British Columbia and vicinity. Pacific halibut are targeted along sandy or muddy bottoms, where rockfish are less likely to be caught but which still have isolated boulders that may snag and damage fishing gear. Along the coast, Pacific halibut are most often found between 30 and 270 meters—a depth range that overlaps with that of long-lived rockfish. In combination, these facts suggest that pre-contact fishers actively fished deep bottom habitats where Pacific halibut could be targeted and the bycatch of long-lived rockfish reduced. Given that Pacific halibut live to only 55 years—and, consequently, are less vulnerable to fisheries than yelloweye and other rockfish that live twice as long—these practices are consistent with active conservation.

Although archaeologists have yet to use DNA to examine the selectivity of ancient rockfish fisheries elsewhere in the northeast Pacific, long-lived rockfishes remained abundant throughout the region into the 20th century. Indigenous people likely employed strategies for the sustainability of rockfish fisheries throughout the entire coast for millennia, even if such strategies differed from those in Barkley Sound.

But not all that starts well stays well. Rockfish overexploitation in British Columbia became rampant between the 1970s and 1990s—a period representing less than 1% of the documented history of Indigenous fisheries for rockfish—when industrial fisheries expanded and removed 90% or more of the biomass of many rockfish species in

British Columbia. Indigenous people and scientists (Indigenous and non, myself included), with valuable input from commercial fishers, are now working together to clean up that mess via a combination of protected areas and more conservative fishery management.

But this history is about much more than rockfishes. It is about a fundamental question that has occupied my mind for many years.

Are humans inherently destructive? If your knee-jerk response is "yes," then chances are that you have been paying attention to trends in our climate, species extinctions, and the loss or degradation of ecosystems that sustain our individual bodies, collective economies, and more intangible values such as inspiration, peace, and our sense of connection and belonging. And we humans are the ones who are causing these losses. So while we are at it, we may as well rephrase the question as Are humans inherently *self*-destructive?

If your answer is still "yes," then you are in the same camp as many environmental scholars. And at first glance, who can argue?

There is no doubt that humans have changed the world irrevocably. The severity of such change is encapsulated in fossils of trilobites—ancient sea creatures related to and reminiscent of today's horseshoe crabs—that had lain buried for the past 507 million years under the rocky bottom of a former sea in what now is the desert in modern-day Utah. Recently, geologists studied these trilobites, searching for clues into their ancient environments and lifestyles by studying biomarkers: substances that are environmental in origin but that infiltrate the living cells, dead remains, or fossils of organisms. But only a fraction of the geologists' findings qualified as ancient. The chemical composition of the half-billion-year-old trilobite specimens is now largely comprised of artificial chemicals that we, humans, began synthesizing in corporate labs less than 100 years ago: plasticizers, flame retardants, petroleum by-products, and insect repellent. Less than half of the biomarkers—41% on average—held traces of microorganisms and plants from a world that preceded the recent explosion of synthetic chemistry. In the blink of an eye, our modern compounds have largely diminished the chemical storytellers of primordial Earth.

The power of the trilobite study is more than symbolic. What we do to the geologic record, we do to our insides. In the words of Sally Walker, a coauthor of the study, "Plastics, pesticides, and other petroleum byproducts are the worst. They affect our endocrine and reproductive systems.... I am sure that my entire body is outlined by plastic from all the plasticized paper I've touched and all the water I've imbibed."

When put that way, then maybe we should cling to the story that humans are inherently destructive and are particularly good at being self-destructive. Tempting. Very tempting. And a total cop-out.

It is true that we, humans, have excelled in our capacity to be a geologic force. In doing so, we have catapulted ourselves out of the Holocene, the 11,700-year-long geological period that began at the end of the last Ice Age and allowed civilization to develop under a stable climate. Those days are now replaced by the Anthropocene: the new geological epoch in which the atmosphere, ecosystems, and geologic materials that comprise Earth reflect the collective actions of dominant human cultures.

Yet it is even more essential to recognize that the human potential for causing further change—positive or negative, constructive or destructive—is still latent. The next many millennia on Earth will reflect the stories that *Homo sapiens*, the tool user, decides to ditch or to accept into our collective identity during the second and third decades of the 21st century. At this pivotal moment in history, the most important story we can be telling ourselves is that humans are *not* self-destructive. This reinterpretation of ourselves is not mere fancy mired in nostalgia for a golden age that never was. Rather, it is consistent with the world views and actions of Indigenous cultures that took shape millennia ago and that are still very much alive, integrating the traditional and the modern, teaching others and learning from others.

As the environmental scholar Jennifer Jacquet reminds us, "Not all humans are a geologic force—and a geologic force is not what humanity must be. That humans have become the main driver of environmental change is largely the result of specific cultures mixing with specific economic systems and mixing with specific technologies."

These ideas ask us all to work. They demand that we let go of the comfort and ease of riding along with the story that humans are hard-wired to destroy. Letting go of that story is difficult. Otherwise, the need to reinterpret ourselves would not be as real and urgent as it is today.

2

Resisting Least Resistance

I had to write to change my story. Not the one of where I came from but the one of where I stood and was willing to go. That is why my first book—a collection of letters to my daughter about the state of the world and our possible futures—took hold of me. It refused to let go until the tumble of words found its own stillness.

I, too, had bought into the story that humans destroy all living things. This convenient ideology granted me the self-justification to detach from deeper inquiry into what humans actually were, or could be. Then I realized that such a negative view of my own species could become my daughter's downfall—and was fast becoming mine.

To be fair, my negativity was not entirely off base. I am an ecologist, active in research and well-versed in the climate change literature. I read and write scientific papers, collect and analyze data, and try to understand how ecosystems respond to the different ways in which people carry themselves in the world, from destructive and unaware to gentle and intentional. That means that, by profession, I am cursed with inside knowledge, in intimate detail, of the damage caused by human actions on the planet.

During a forest walk amidst centuries-old Douglas fir trees, I cannot ignore the exotic plants—which humans have thoughtlessly brought from other continents in recent centuries—that have invaded the understory and displaced native species. While scuba diving among rocky reefs, I am haunted by the absence of large predatory fishes: an ecological vacuum created by industrial fishers. During a mountain hike, an open view of the sky triggers thoughts about how our fossil fuel emissions have altered the chemical composition of the atmosphere for centuries to come, with dismal repercussions for all living things.

But that is only one side of my story. By profession, I also carry the gift of seeing the living world in its wondrous and interconnecting details. To show you what I mean, let me take you back to a recent September morning, near my home on the temperate coast of British Columbia, Canada. I was out on the ocean in my kayak. It was a day off, and I was allowing my curiosity to chart its own course. I drifted towards an immature turkey vulture. The big black bird stood, immobile and stately, at the edge of the rocky shore. It stared past me, out towards the sea where the tide was beginning to run against a slight southerly wind, stacking water into steeper waves.

The vulture's black head and face were smooth and iridescent. Its head—featherless to allow for plunging head-first into dead mammals' body cavities to feed on entrails—reflected the morning light. My eyes and nose searched the rocks for signs of a dead seal or sea lion that might have attracted the bird. I found none. Yet the sheer act of looking made me wonder how scavengers, like this turkey vulture, might transport energy and nutrients between ecosystems.

My mind began to connect images that other ecologists had captured, through careful experiments, in their own drive to understand natural mysteries. Herring schools leaving hydrodynamic trails—the liquid equivalent of scent, expressed as subtle-yet-lingering turbulence in the water. Seals and sea lions immersed in the pitch-dark of night using their whiskers as sensors to follow those trails and feed on those herring.

Then I remembered that, a few years back and close to this spot, I had seen a massive California bull sea lion dead on the rocks, half a dozen vultures sticking their heads into its body. All of this made me wonder whether vultures feeding on dead seals and sea lions, and nesting on forested hills like the one behind my house, might fertilize through their feces upland trees and shrubs with nutrients that originated in herring. This idea, I realized, could be tested using stable isotopes—the same chemical technique that helped ecologists understand that bears fertilize the forest with the carcasses of salmon captured in streams and dragged into the bush.

The thought was exciting, and I really needed that. You see, the sea surface temperature that day was 21°C: unusually high for the season and an obvious symptom of climate change. By focusing on the natural history around me, I could focus on the pulsing, living planet that we still have, rather than on the pessimistic scenarios that hover in the shadows of many of my workdays.

Days in which reconnecting with the Earth boosts our motivation to keep going are not unusual in my own life and that of my colleagues. Such days keep us from going crazy as we grapple with climate change and the disruption of ecosystems caused by, yes, humans. Our lived experience confirms that bad news may be a part of modern ecology, but so are wonder and the possibility of good news. British entomologist Miriam Rothschild said it best: "For someone studying natural history, life can never be long enough." That is as true today as it was millennia ago, when humans first learned that tracks on the ground told the stories of animals, that insects and birds pollinate the flowers that yield delicious fruits.

The birth of my daughter, in 2004, thrust upon me a dual task: to be scientifically realistic about all the difficult changes that are here to stay, while staying humanly optimistic about the better things that we still have.

By the time my daughter turned eleven, I had jettisoned my nostalgia for the Earth I was born into in the mid-1960s—a planet that, of course, was an ecological shadow of Earth 100 years before, which in turn was an ecological shadow of an earlier Earth. The pragmatist in me had embraced the Anthropocene, in which humans dominate all biophysical processes, and I ended up feeling genuinely good about some of the possible futures in which my daughter's generation might grow old.

It was a choice to engage in a tough situation. An acknowledgement of rapid and uninvited change. A reaffirmed commitment to everything I have learned, and continue to learn, as an ecologist working with Indigenous people on marine conservation. Fundamental to this perspective is the notion of resilience: the ability of someone or something—a culture, an ecosystem, an economy, a person—to absorb shocks yet still maintain their essence.

But what is essence?

Ecologists have one kind of answer to this question. In the early 1970s, Buzz Holling—a highly influential ecologist and, I like to boast, my academic grandparent (Buzz supervised the dissertation of another influential ecologist, Larry Dill, who in turn supervised my dissertation)—formalized the concept of ecological resilience into the language of Western science. The notion was already well understood in the traditional knowledge of many Indigenous cultures, yet Buzz pioneered its mathematical expression and was the first to acknowledge its implications for how industrialized societies exploit ecosystems. Generating beautiful curves, concentric circles and spirals, he described how external forces disturb ecosystems—forests burn or are logged, wetlands flood or dry out, exotic species infiltrate food webs. By itself, that was old news. But Buzz went further by explaining how, despite these disturbances, ecosystems could maintain their essential structure and function, as long as certain interactions and processes endured. For a forest, that might mean being burnt to a crisp yet not losing its ability to regenerate over time into a vertical woody structure resembling that which existed before.

History, however, is unlikely to repeat itself exactly. Extinctions, invasive species, and climate change might preclude the regenerating forest from maturing into the exact same combination of species or age structure that it held in the past. Yet a shift in species composition does not necessarily change the essence of an ecosystem. In the case of the forest, this essence includes slugs or other decomposers breaking down plant matter so that the nutrients are incorporated back into the soil; trees providing structures for lichens, mosses, ferns and cavity nesters; predators like wolves and cougars keeping herbivores like deer from overbrowsing plants; high winds breaking branches and knocking down individual trees, thus creating gaps in the canopy that allow light to pour in and support a vibrant understory.

The idea of resilience, and its explicit acknowledgment of change, is very different from the more old-fashioned (and unrealistic) view of Western conservation, in which ecosystems are intended to stay as they were when we first became aware of them. We must not, however, misconstrue resilience thinking as licence to disturb everything in the name of human business. Crossing a threshold of disturbance will hurl ecosystems into something fundamentally different, from which they are unlikely to come back—what ecologists call an alternative stable state—such as a forest pushed beyond the possibility of regeneration becoming a biologically depauperate shrubland. More resilient ecosystems can take in more shocks without being thrown into an alternative stable state.

These ecological notions helped me make my peace with the Anthropocene. They gave me tools for distinguishing human-caused changes that I can live with (those that do not reduce resilience) from those that I fiercely oppose (those that, alone or combination with other stressors, erode resilience). Effectively, this means managing humans so that whichever disturbances we *can* control—how much we log, how much we fish, how much carbon dioxide we release into the atmosphere—do not create the perfect storms that jolt ecosystems into alternative stable states. And when alternative stable states do become inevitable, as they certainly will in our changing climate, resilience thinking might help us steer towards the ones that are more benign—something akin to anticipating engine failure in a plane yet being ready to make as smooth a landing as possible.

I constantly think about the parallels between ecological resilience and Indigenous cultural resilience. Part of it is intellectual curiosity. But mostly, I am just trying to hand my daughter's generation a better world.

Indigenous cultures that derive their resources and world views from the myriad wild species that they hunt, fish, gather, and tend have existed for thousands of years. Over time, as resource managers, they have developed intentional and socially complex practices that epitomize sustainability. Some of these cultures were crushed by colonization. Yet many, despite immense historical challenges, endure: they maintain their essence while adapting to a rapidly changing world. Their experiences are not devoid of severe loss, yet they also embody resistance, intentionality, self-identity, and resilience. Perhaps they provide a model for how we might navigate the irrevocable changes we have unleashed upon our planet and help us steer towards a more benign Anthropocene.

Through my day job, I regularly spend time with some of the Indigenous leaders who are helping their communities emerge from the dark horrors of colonization, revitalizing their cultures into something that is both modern and grounded in tradition. I have stacks of handwritten journals from my time in villages and in the field, and those journals, I believe, hold key insights into the issues that I seek to understand.

Yet those same journals also throw hard questions at me.

Human impact on the planet precedes industrial civilization by a long shot. The collective psyche of agricultural civilization began to break apart many ecosystems about 10,000 years ago, triggering a barrage of extinctions, both biological and cultural, as large settlements depending on a very small number of cultivated and farm-raised species began to displace complex ecosystems and the peoples who depended on them. There also exists little doubt that, thousands of years before the first harvest of domesticated grains ever occurred, the collective psyche of very early hunter-gatherers, as they first arrived on continents previously uninhabited by humans, contributed to massive extinctions: notably in Australia 55,000 to 42,000 years ago and in the Americas 16,000 to 14,000 years ago. A changing climate likely made many species more vulnerable to overhunting, but that does not diminish the likelihood that, in the absence of humans, fewer extinctions would have occurred.

But no matter how you look at it, as far as destruction goes, industrial civilization has far outperformed preindustrial agriculturists and prehistoric hunter-gatherers. Just in the past century, we have raised the Earth's average temperature by 1°C, and we are still cranking up the heat, propelling the planet into a new geologic era that already has caused many species extinctions and is altering the future of all ecosystems for millennia to come. The scales of ecological disruptions caused by preindustrial agriculturists and prehistoric hunter-gatherers were child's play by comparison.

Recognizing the collateral damage of the Agricultural and Industrial Revolutions has been straightforward for me. The evidence for these historical landmarks is crystal clear. And maybe I want to believe that these have been our species' teenage stages, when the societal equivalent of an underdeveloped neocortex has kept us from fully understanding the consequences of our actions.

But it was more difficult for me to reconcile the deep past, when prehistoric hunter-gatherers did cause many extinctions, with what the human capacity to conserve might actually be. Are humans really that innately destructive?

Maybe, maybe not.

An alternative view is that humans *have been* "deadliest" only during specific historical periods or when belonging to specific cultures. That is, prehistoric hunters wiping out entire ungulate herds, agriculturists running amok, and climate-altering industrialists represent incomplete windows into the ways many human cultures have carried, and still carry themselves, in the world.

Alternative interpretations of human history give greater emphasis, and credence, to the world views of cultures that connect strongly to the specific places where they originated and still exist, and that have developed reciprocal relationships with the native species that share those places. My short version of that reinterpretation is as follows:

After some centuries of raising ecological havoc in their "new" continents, hunter-gatherers settled down. The land ceased to be an expanding frontier and became home, the place where we humans can find our medicines, our songs, our stories—the reasons why we exist at all. Once committed to staying in place until the end of time, they became invested in their descendants. They continued to hunt, fish, and gather plants, yet they stopped killing and consuming resources anywhere and anytime. Instead, they began to actively manage themselves, curtailing waste and tending the landscape in ways that promote the diversity of native species. Had they not changed their ways, they, too, would have become extinct. Once rooted, many cultures became visionary in their long-term resource management strategies—identifying sustainable rates of harvest for different species and transmitting this understanding intergenerationally—to the point that human-caused extinctions ceased in their continents until the arrival of Europeans.

Granted, the Indigenous picture is not uniform. A mere few centuries ago some societies that developed in the new continents caused their own ecological havoc that led to social collapse. But these collapses involved agricultural societies, such as the Maya and Anasazi of Central and North America, who had lesser connections to the broader diversity of species and instead simplified ecosystems in ways that favored a few domesticated plants and animals.

In contrast, cultures that remained connected to the larger variety of native plants, animals and other organisms for their fundamental economics and spirituality—underlying their connections to biodiversity with reciprocity and respect—continued to thrive, and likely would have continued to do so had they not been severely disrupted by European colonization over the last couple centuries. As I will elaborate over the course of this book, these cultures are still here, revitalizing themselves and increasing their political influence in some countries, such as Canada. Even so, their legacy and modern influence remains, so far, dangerously underappreciated. To ignore the accomplishments of cultures whose very existence depends on reciprocal and respectful relationships between themselves and their ecosystems is to shortchange the human potential.

Sixty-six million years ago a huge asteroid collided with Earth. The shock unleashed earthquakes, tsunamis, and volcanic eruptions, obliterating three quarters of animal species on land and sea. I like to think of the emergence of some human cultures that crystalized much later, only within the last 14,000 years or so, as the inverse of that event. Wherever people ceased to live like roving bandits, sequentially depleting area after area, and instead became inherent to specific places and to their functioning ecosystems, biodiversity thrived and persisted.

Dúqvạ̊ísḷa (William Housty), a Heiltsuk hereditary chief dedicated to integrating the traditional knowledge of his people with scientific methods for resource management and conservation, points out that, according to customary laws known as *Gvi'ilas*, "Heiltsuk have been present in traditional territory since time began and will be present until time ends." This belief, which is shared by all First Nations with whom I have spent time in Canada—both coastal and inland—arguably is the most powerful conservation tool imaginable.

The moment you believe—in your spirit, your gut, your whole being—that the surrounding lands and waters are where your people have lived "since time began," and that those very same places are where your people will *stay* "until time ends," a cascade of commitments and responsibilities begins to flow.

This cascade, Dúqvạ̊ísḷa writes in reference to *Gvi'ilas*, includes "responsibility over traditional territory as much as over immediate home," which entails that, "out of respect and understanding, certain areas should be off-limits to some, or all, human activities." Further, "The right to use a river system [or any other part of the land or sea] comes with the responsibility to maintain [it] in its natural or ecological entirety." And critically, the focus of resource management decisions "should be on what is left behind, not what is taken."

But what about those of us who are *not* Indigenous? That includes me, the descendant of immigrant Jews who centuries ago left the Middle East, scattered over eastern Europe and Mediterranean countries and, more recently, barely escaped the Holocaust as they found sanctuary in Mexico. There, only two generations were born before my immediate family migrated once again, this time into Canada. I, who descend from wandering ancestors, am the antithesis of Indigenous. Does that mean that the notion—and conservation power—of becoming part of a place and its functioning ecosystems is unavailable to the likes of me? I sure hope not.

Indigenous peoples currently are only 5% of the global population. If the remaining 95% of humans lack connection to place, chances are that future Earth will be grim.

So how do we become part of the land on which we stand? Even if settler families arrived at their present countries several generations ago, it takes intentionality to become inseparable—spiritually, physically, viscerally—from the biodiversity that defines a coastline, a river valley, a forest, or any other landscape. Culture creates that intentionality, yet cultural appropriation does more harm than good. Becoming Indigenous, genuinely linked to a place "since time began," is off the table for most of us. Yet disconnection to place is *not* the only alternative.

Robin Wall Kimmerer, a member of the Citizen Potawatomi Nation and a plant ecologist at the State University of New York, has thought hard about these issues. Her positionality as an Indigenous woman and an accomplished member of the global scientific community has led her to powerful insights. And to answer some of the hardest questions facing humanity, she often turns to the wisdom of plants. Of invasive species brought by European settlers into North America, she wrote,

> Garlic mustard poisons the soil so that native species will die. Tamarisk uses up all the water. Foreign invaders like loosestrife, kudzu and cheat grass have the colonizing habit of taking over other's homes and growing without regard to limits.

These bullies, which spread with impunity and simplify ecosystems into low diversity and low resilience, embody only one possible narrative for what a settler can be, and often *has* been. But, Kimmerer points out, settlers also can access an alternative narrative: the one symbolized by plantain, *Plantago major*, which "fits into small places" and "coexists with others."

> Just a low circle of leaves, pressed closed to the ground with no stem to speak of, it arrived with the first settlers and followed them everywhere they went. It trotted along paths through the woods, along wagon roads and railroads, like a faithful dog so as to be near them.... Its Latin epithet *Plantago* refers to the sole of a foot.

Kimmerer goes on to describe how her people first distrusted this "White Man's Footstep" but eventually "became glad for its constant presence" as the seeds "are good medicine for digestion. The leaves can halt bleeding right away and heal wounds without infection."

Kimmerer's insight is that plantain is "so well integrated that we think of it as native." Yet it is not. Plantain is *naturalized*: the term used for immigrants into a new country who "pledge to uphold the laws" of that specific part of the world.

The First Nations with whom I work—Nuxalk, Kitasoo/Xai'xais, Wuikinuxv, and Heiltsuk—have ancient and sophisticated legal systems that outline human responsibilities towards human and non-human kin. Paramount to these laws are principles of respect, gratitude, and reciprocity towards all living things. Of taking only what we need, not wasting. Of making management decisions that prioritize what is left behind rather than what is taken. These laws are codified in traditional stories and in the archetypal symbols used to create art. When Europeans arrived at the Americas, many other cultures, including Kimmerer's people, practiced similar legal systems. For the First Nations who I work with, as for many other Indigenous Peoples, these traditional legal systems are still very much alive and being incorporated into modern resource management (a theme I will return to in later chapters).

So there you have it.

The likes of me will never become Indigenous; nor should we want to. But we *can* become naturalized, mindful and respectful of Indigenous laws. Those that, in the tradition of Kimmerer's people, provide the "Original Instructions" for practicing an "Honorable Harvest," one that promotes the diversity of life and the resilience of ecosystems.

Following the paths of kudzu, loosestrife, and cheat grass is a choice. But so is a future shaped by Kimmerer's alternative narrative. *Plantago major*, a naturalized citizen who came from faraway to practice coexistence, to share healing powers that benefit settler and Indigenous both.

3

Coalescing Knowledge

For six years and counting, my work as an ecologist has focused on marine conservation in collaboration with modern Indigenous people. Because of my day job, I stand—intellectually, emotionally and spiritually—at the crossroads of history, culture, ecology, and social justice.

For centuries, and in most parts of the world, Indigenous cultures with reciprocal relationships to biodiversity—whose very existence depends on sustainable tending to and fishing, hunting, and gathering of native species—have been kicked hard by colonists who come from faraway. Much of the damage has been done by European settlers driven to homogenize the world into the ways of Christianity—where food consists of a tiny number of domesticated species that sprout from rows in the ground (wheat and a few others) or stand semi-immobile on four legs, burping methane and posing to be tipped-over by teenagers in the middle of the night (cows and some of their close relatives). Sure, a few other types of domesticated species have been involved (chickens, pigs, goats); the so-called frontier expansion has not always been European or Christian (though almost always monotheistic); and the pursuits of fossil fuels, timber, minerals, and other resources have also played a role. Yet, no matter the slight variants in its motivation, colonial mentality has thrived on simplifying the world. Colonists must eliminate, or at least severely diminish, native species and Indigenous cultures that might obstruct "civilized" uses of the land over which they believe they have been granted divine authority. The consistent outcome has been a suppression of world views about how humans might interact with all other living things, a narrowing of our collective psyche.

This kind of history applies to the Indigenous peoples of Canada, including the First Nations with whom I work. For over 150 years, a policy of forced assimilation became the government's solution to the so-called "Indian problem." The Canadian government forcibly relocated Indigenous people away from traditional food sources that are fundamental to their cultural identity and into settlements with ramshackle houses that lacked running water. Authorities muted original languages and criminalized ceremonies and dances. Between 1884 and 1996, government and church agents stole 150,000 Indigenous children from their families and confined them in residential schools, where they were abused and dehumanized. All the while, diseases introduced by the colonists decimated the original inhabitants of the land.

When in 2015 my colleague, Lauren Eckert, interviewed people from Central Coast communities about traditional harvesting and stewardship principles learned from family members, one person answered, "No, my father never brought me up; the residential school brought me up." And a younger adult explained,

> I remember elders, when I was younger.... The only language was the native [traditional] language. So, I remember some elders trying to tell stories, but they would need someone to translate for them.... They said the words [between the two languages] were so different that a lot of the stories were lost through translation. [The next] generation, [many of them] can't tell me a story, can't tell me a song or dance.... And I think the products of residential school didn't really teach them the ways.... None of [the language, culture, or stories] were passed on to us as kids, and that was really unfortunate but that's just the way it was....

Remarkably, some of the Christianity's tactics for cultural annihilation—carried out within living memory—resemble those used during the Medieval Inquisitions in Europe. When asked about traditional stories, an elder replied,

> Christianity was really strong. They didn't allow anyone to pass a story to us.... There were a lot of masks, totem poles, talking sticks, rattles; they were all here! They were gathered and burned by the Church. Now we're slowly getting it back.

It is tempting to hang onto the image of that violent inferno. Instead, the people I work with focus on the elder's final words: "Now we're slowly getting it back."

An explicit part of my job with the Wuikinuxv, Kitasoo/Xai'xais, Nuxalk, and Heiltsuk Nations is to collaborate with them in the integration of traditional and scientific ways of knowing, and to apply the results to the conservation and management of marine ecosystems. As applied to this work, traditional knowledge is the understanding of ecosystems accumulated intergenerationally by place-based peoples within their specific cultural context and belief system. Traditional knowledge often is supplemented by local knowledge: contemporary observations made by individuals who are keenly aware of their environments and trends in the resources that they harvest. In addition to being holders and practitioners of traditional and local knowledge, many Indigenous individuals are ecologists or resource managers who apply the tenets and tools of Western science.

Traditional and scientific ways of knowing can merge in many contexts, helping people find better ways to manage their relationships to local ecosystems. Yet for the two knowledge systems to coalesce most effectively, it is important to understand their commonalities and differences, and how these may be complementary.

Both ways of knowing seek to understand forces—predation,

exploitation by humans, nutrient flows, tides, climate, and others—that affect the abundance and distribution of plants, seaweed, animals, and other organisms. In doing so, both ways of knowing strive to predict changes in biological communities in response to human behavior or other dynamic actors.

Ecologists capture their predictions in theory, often invoking mathematics, while keepers of traditional knowledge capture their predictions in laws and stories, sometimes invoking the supernatural. Scientists like myself may consider the supernatural a metaphor, perhaps loosely analogous to mathematics in the sense that both can encapsulate an interpretation of ecosystem dynamics. For instance, when a traditional story describes humans transforming into animals and vice versa, there is a message about shared use of resources and interdependence between humans and other species. Ecological theory can describe the same relationship, albeit in a different way and, depending on the world view of the theorist, perhaps with a different emphasis.

But that is only my view: one that belongs to someone raised outside an Indigenous tradition. Still, that perspective has helped me make sense of the knowledge that has been codified into the laws and stories that have been shared with me. It has illuminated for me that both traditional ecological knowledge and ecological science recognize the importance of holism and synergism—the interconnection of all living and physical entities—and that some combinations of species may have a larger role than others in maintaining the resilience of ecosystems. Old-school fisheries science, in contrast, focuses on one commercially valuable species at a time, which makes it obvious to many of us why so many fish stocks have been mismanaged into collapse and why marine ecosystems have degraded throughout most of the world.

Both ecologists and holders of traditional knowledge rely on cumulative and collective observations that are transmitted intergenerationally. Ecologists train with mentors, read texts by pioneers of their subdiscipline that may have died a long time ago, build contemporary thoughts onto key ideas conceived decades earlier, teach students (and vice versa), and disseminate our findings in the literature or through lectures so that everyone else can access (and criticize in constructive discourse) what we learned. That is, for the most part, science advances incrementally: younger generations listen to their scientific elders and pick things up where the earlier momentum might have stopped. Somewhat similarly, traditional knowledge is shared and expanded across generations, yet a key difference is that it often has an explicit set of proprietors who may choose to restrict that knowledge to a specific inner circle.

Another distinction is that the lineage of people who hold traditional knowledge goes back in time for millennia, while the scientific method has been formalized for a mere 500 years or so. Some might say that this difference is trivial because, when it comes to knowledge itself, scientific inferences appear to address longer time frames than traditional world views can. After all, cosmologists estimate the universe to be nearly 14 billion years old, and biologists date the origin of life on Earth to nearly 4 billion years ago. Some Indigenous stories, however, speak of a time before time which is arguably analogous to a time before the Big Bang; interpretations about who has captured the longer time frame may reflect the cultural norms of the interpreter.

Traditional knowledge, however, might be less likely to recognize ecological phenomena that we cannot observe directly. What prompts me to say this is an increasing awareness by scientists of the diversity and ecological importance of microbes. These single-celled organisms are invisible to the human eye and include bacteria and protists. (Protists have a cellular nucleus while bacteria do not.) Thanks to recent advances in biogeochemistry and microbiology, we now know that microbes make up a very large proportion of living matter in the ocean and play key roles in recycling the nutrients that keep ecosystems ticking. I am unaware of traditional knowledge capturing the ecological role of marine microbes. But then again, I might be thinking too narrowly. Perhaps traditional stories already capture the complexity and importance of microbe diversity in some metaphorical way, and people like me have yet to listen closely enough to get it.

Another difference is that traditional knowledge often focuses on the geography of its particular culture. In contrast, ecological studies can teach any of us about ecosystems that are far away and that we are unlikely to ever visit in person, such as hydrothermal vents kilometers beneath the surface of the ocean or underwater ecosystems of Antarctica. Sure, "book" knowledge is not the same as lived experience, but it is knowledge all the same. Practitioners of Western science (Indigenous or otherwise) have access to satellites, deep sea probes, and other computerized gadgetry. While the adaptability of Indigenous cultures does not preclude using such tools to expand traditional knowledge, it is the scientific context in which these tools were developed that has led to the measurement, visualization, and prediction of phenomena for which there is no precedent in human history. These phenomena include the rapid rates of ocean acidification, species extinctions, and climate change we are experiencing globally.

But, despite all their nifty technological tricks and access to "book" knowledge from faraway places, scientists often compromise themselves by being perhaps too mobile in their personal and professional lives. Sometimes because of a legitimate need to chase job opportunities, but often to enact the belief that gallivanting all over the globe expands your intellectual views, many ecologists jump ship from one geography to another on a regular basis. If you do that a lot, you do glimpse many ideas that originate in different parts of the world, but you may also become less likely to ever connect, viscerally, with any particular place. Such mobility limits the types of questions scientists might ask about how place-based peoples interact with their environment. This criticism does not apply to all scientists, especially not to those who are Indigenous and remain connected to their cultures, but it does apply to many. So, unless coupled with traditional and local knowledge, science may yield only limited insight into certain types of ecological problems. In contrast, traditional knowledge epitomizes direct connection to the resources that provide physical and spiritual sustenance and ancestral rooting. That means that place-based Indigenous peoples are more invested in those resources, and in the long-term knowledge of those resources, than most scientists will ever fathom. And oral traditions preserve that knowledge as laws and stories that transcend many limitations of science.

Both scientists and practitioners of traditional knowledge are empiricists. What they understand with confidence derives from what they observe. But the two arguably differ in their degree of empiricism. If a notion that potentially explains a phenomenon cannot be tested in a way that might allow us to reject it, then that explanation is generally considered to be outside the realm of science. So cosmologists test ideas about the age and expansion of the universe by observing light in space and testing predictions derived from physical principles about visible wave lengths. In contrast, the idea of Raven bringing light to the universe for the first time, as depicted in the oral tradition of the Haida Nation, is not falsifiable. But then again, science is science and traditional knowledge is another way of knowing; the two, despite being complementary, are not equivalent.

When you put it all together, practitioners of traditional and scientific ways of understanding ecosystems can complement each other quite productively. The potential result is a stronger foundation for conservation and resource management policies. Or, more to the point, a model for how we might change, fundamentally, the collective psyche of industrial civilization.

Although I believe that the synergy of these two ways of knowing is essential to solving some of the biggest problems of our times, traditional ecological knowledge has the stronger track record for systematic conservation planning and achievement. I will illustrate this point with the case study I know best: that of Coastal First Nations of British Columbia who, for thousands of years prior to European col-

allowed some wiggle room for human populations to grow without overexploiting their environment.

Still. Five to ten millennia provide plenty of time to mismanage and deplete resources, especially because human populations were large in the Central Coast, as evidenced by the large number of village sites and high densities of dwellings and middens within those sites. Yet Coastal First Nations appear to have not taken for granted their good fortune of living in a biologically rich place. They had the power to overexploit resources in ways that would have led to their eventual depletion, yet apparently they chose not to.

As you may recall, archaeological evidence suggests that Indigenous fisheries for rockfish in Barkley Sound, British Columbia, selectively targeted shorter-lived species over the course of 2,500 years. Species with potential life-spans of a century or more, which are extremely vulnerable to overfishing, appear to have been fished infrequently. Such selectivity was made possible by the fact that these species differ in their use of the water column: the longer-lived species live near or just above the rocky reefs, while the shorter-lived ones spend more time swimming mid-water and far above the bottom. These differences in use of vertical space appear to have allowed Indigenous fishers to sink their fishing gear partway into the water column, avoiding the bottom to selectively catch shorter-lived species. Given that longer-lived rockfishes are excellent food and would have been caught easily by fishing technologies of the deep past, going easy on these species appears to have been an active choice.

And, to my mind, a powerful example of the human capacity to conserve.

Rockfishes are only a small element in a much bigger constellation of reciprocal relationships between people and other species on the Central Coast. Pacific salmon, eulachon, and Pacific herring, for example, are three types of fish highly prized by Indigenous peoples, both historically and currently. Though biologically different in many ways, they share a key life history characteristic that shapes much of their relationship with fishers: salmon, eulachon, and herring spend most of their lives offshore, where they rove and are difficult to find. Mature fish of these species, however, aggregate seasonally to spawn, either in nearshore areas of the ocean (as herring do) or in rivers (as eulachon and salmon do), where people can encounter them more predictably and in large numbers. Salmon and herring are the two marine species most widely used, historically, by Indigenous Peoples of British Columbia and adjacent areas of Southeast Alaska, Washington, and Oregon, where archaeologists have found their remains in nearly all middens examined so far. Eulachon, a type of smelt that is extremely fatty, have a much narrower spawning distribution than herring and salmon. Their remains are not as ubiquitous, yet for Indigenous cultures with historical access to rivers where eulachon spawn, including the Wuikinuxv and Nuxalk Nations, the cultural significance of eulachon as a diet item, object of trade, and subject of stories is no less important than that of salmon or herring.

Indigenous fishers have been restraining their capacity to overharvest aggregations of spawning fish for a long time. Archaeological evidence suggests that continuous use of Pacific salmon at the ancient village of Namu, in British Columbia's Central Coast, lasted about 7,000 years, until the village was abandoned in the aftermath of colonization. The longevity of salmon fisheries by the original inhabitants of Namu was possible only because people recognized that repeated overharvesting of spawning aggregations would eventually deplete local stocks. Again, this does not reflect any kind of technological limitation. A combination of archaeological data and oral histories suggest that stone wall traps—often more than 100 meters long—were open in intertidal estuaries during a falling tide to capture fish about to migrate upstream. The traps served as holding pens where fish could be gaffed, harvested selectively according to species or size. Critically, these traps and similar technologies, such as weirs, were deployed only long enough to harvest what was needed and then deactivated preemptively to preclude overexploitation.

First Nations acquired the ability to store dried fish for the lean winter months thousands of year ago, which introduced new challenges for conservation. Food preservation allows people to harvest much more than can be eaten before spoilage, and it therefore amplifies the risk of overexploitation. The black leather chiton, a large mollusk found on intertidal rocks and priced by the Sugpiaq people of western Alaska, provides a modern example. Electricity and freezers were introduced to Sugpiaq villages in the 1970s, and harvesting protocols did not immediately compensate for the new technology. In the words of a traditional harvester speaking to ecologist Anne Salomon, "In the past we picked just enough to eat and snack on. But when electricity and then freezers came to the village, people began to pick more because they could store them," and these changes contributed to local declines of black leather chitons. It is possible that similar declines of salmon and other species occurred in the Central Coast when First Nations originally mastered the ability to dry fish. But the fact that salmon fisheries remained sustainable in Namu for as long as seven millennia suggests that fishers eventually learned to avoid resource depletion by harvesting only enough fish to meet their short-term needs for fresh food and their longer-term needs for feeding themselves during the winter, before seasonal harvests could resume.

Conservative management principles also apply to the Indigenous relationship with herring, eulachon, clams, crabs, seaweed, sea urchins, blueberries, cedar trees, and myriad other species. This resource portfolio is managed holistically, which requires people to be aware of the state of multiple species simultaneously, and to intentionally switch between species at different times so that none are overharvested. The archeological record indicates that less-prized-yet-still-important foods—including smaller, leaner, and bonier fish like surfperches, sculpins and greenlings—were consumed often, perhaps as a way of relieving pressure on the choicest species.

In contrast to the Indigenous portfolio approach, modern industrial fishers have a track record of targeting the best and easiest pickings first—whatever is closest, shallowest and has the best market price—until the near-annihilation of a resource forces them to go farther and deeper in pursuit of other species that might yield a lower financial profit. That *modus operandi*, known as serial depletion, has thrived globally under perverse government subsidies, altering the oceans in irrevocable ways over the past 100 years.

Yelloweye rockfish are an example close to home. Over their lifetimes, fishers from the Nuxalk, Kitasoo/Xai'xais, Wuikinuxv, and Heiltsuk Nations have watched the average body length of yelloweye in their catches decline from more than 80 centimeters between the 1950s and the 1980s to nearly half that size in recent decades. The shrinking of yelloweye and other species coincides with the explosion of industrial fishing that occurred during the late 1970s and 1980s. Despite tremendous improvements to the management of industrial fisheries that began in the 2000s, the situation remains worrisome.

I often work on projects with my colleague Madeleine McGreer—a talented statistician and the fisheries coordinator for the Central Coast Indigenous Resource Alliance—and recently we examined data from over a decade of research surveys. The results showed a fast pace of ongoing declines in the size and age of long-lived rockfish within the territories of Central Coast First Nations. Between 2003 and 2015, the average age of yelloweye rockfish declined by nearly 10 months each year. During a similar period, the average quillback and yelloweye rockfishes shrank in length by nearly half a centimeter each year. Smaller, younger females are much less fecund than larger, older females, so these declines signal a loss in the ability of the population to recover from fishery exploitation and stresses caused by climate change.

Globally, industrial fishers have been running out of fishable spaces and depths. Forced into a corner, only recently have they started to pursue less destructive ways of conducting business. "Better late than never," you might say, and rightly so. Mainstream society has a long history of valuing the impunity of markets over everything else and of being blinded by racism to the value of traditional knowledge. Otherwise industrial fishers would have avoided this mess by visiting their First Nation neighbors for some friendly advice that would have kept them from nearly depleting the oceans. But the important point is that people *can* change. And, given the precarious state of our planet, "better late than never" is not entirely vacuous.

Developing the restraint and intentionality with which Coastal First Nations have conducted themselves in the world requires strong cultural norms that infuse the collective psyche of an entire society with a commitment for responsibility towards other species. These norms are codified into stories in which human and non-human worlds have fluid boundaries, intertwined by the powers of transformation.

Traditional stories encapsulating the consequences of failing to abide by the principles of reciprocity and respect are being applied to modern fisheries management. For example, the *Kitasoo/Xai'xais Management Plan for Pacific Herring*, which outlines how herring fisheries are to be managed by traditional laws integrated with modern science, is up front about the consequences of failing to respect other beings:

> Wrongful actions have natural and spiritual consequences. For example, in the story Gunarhnesemgyet seal hunters are out in their canoe when the steersman notices a seal in the mouth of a sea anemone. He urges the hunters to take the seal, but they respond with taunts and scorns, scoffing that they can hunt their own seals without assistance from a lowly sea anemone. Because of disrespect to the anemone, the group becomes stranded and everyone but the steersman dies. Another story recounts a group of young boys on their canoe hooking sea cucumbers for amusement. One of the boys tells the others to stop, but they do not. Similar to the group of seal hunters, the boys are caught in a freak storm and capsize; only one boy survives so that others may learn from his testimony.

While both reciprocity and respect operate at spiritual and legal levels that govern human ethics and behavior, reciprocity stands out for its tangible marks on the landscape in the form of tended landscapes that enhance wild species, allowing them to thrive in ways that are completely unlike industrial agriculture and its objective of homogenizing the landscape into a few domesticated crops.

Tended *and* wild. These two concepts might seem like contradictions to the dualistic mind-sets that created most of our global problems to begin with—"human or natural," "black or white," "with us or against us," "environment or economy"—but not to Indigenous perspectives. Traditional management of the landscape generally fosters, rather than simplifies, the diversity of native species and reflects the permeability of boundaries between human and non-human worlds that are foundational to many Indigenous world views.

Consider clam gardens: rock walls constructed in the low intertidal zone of tidal flats that have been intentionally cleared of boulders and modified to have a gentler slope. This practice is at least 3,500 years old. It reduces the time beaches are exposed at low tide, which boosts the growth rates and overall productivity of clams harvested for food. The rock wall itself also enhances biodiversity by creating habitats for sea cucumbers, crabs, sculpins, and other species.

As for other resources, tended foods were part of the broader portfolio of Coastal First Nations. That is, clam gardens, stone wall traps for salmon, and estuarine root and berry gardens (small patches of shoreline modified to boost the productivity of terrestrial food plants) were built near each other—managed and operated as components of an integrated system.

An awareness of long-term connection to very specific places for resource gathering is so fundamental to the world view of Coastal First Nations and other Indigenous peoples that it pervades everyday language. As one of many possible examples, consider the Nuxalk word *ksnmsta*, which means "our physical food supply and the places where our food comes from, as well as the livelihood that comes from those places." That single word captures a unified notion that English speakers must split into three separate concepts (food, place, livelihood) and invoke many more words to describe. Arguably, the two different languages reflect the contrasting ways in which Indigenous and settler cultures have approached their long-term relationships to the land and sea. Without an explicit awareness of *ksnmsta*, it would have been impossible to harvest resources sustainably for thousands of years.

From the practical standpoint of resource management, all of this is part of the hereditary chief system, a form of marine tenure in which specific people within a lineage inherit the rights and responsibilities to specific areas and their aquatic resources. (Although male chiefs tend to be the norm, gender does not constrain eligibility. For example, a Nuxalkmc woman, Staltmc Q'umulhla (Rhonda Schooner), is one of the current hereditary chiefs of her Nation.) Hereditary chiefs are responsible for when, where, how, and who gathers resources in their chiefdom, and whether such harvests may extend to people of other lineages. Yet this authority is contingent on the chief maintaining knowledge of those resources and their sustainable management, transmitting that knowledge intergenerationally and redistributing his or her own wealth derived from those resources in potlatch ceremonies—gift-giving feasts that provide a cultural foundation for many aspects of governance and resource management. And staying connected to place, for the long run, is a key responsibility of the chiefs. In fact, the hereditary chief system is one of the reasons why, over the centuries and until disrupted by colonization, the Heiltsuk, Nuxalk, Wuikinuxv, Kitasoo/Xai'xais, and many other First Nations were able to live sustainably in place, despite having large populations and sophisticated technologies for resource gathering.

So where do we stand today?

One answer is encapsulated by what Pam Wilson, a Heiltsuk woman about my age, told me in 2015. When she was growing up and people spoke of traditional stewardship principles, she would be told, "This is who we were." Today, when her children are taught about those principles, at school and in the broader community, they are told, "This is who we are."

Much knowledge and practice have been retained. At the same time, some practices have been set aside, at least temporarily. For example, a large part of the portfolio of marine foods is no longer used. None of the people I know from the Central Coast are keen to eat sculpins, greenlings, or shorter-lived species of rockfish that were common food before colonization. When I studied the fishery for halibut, fishers released the bycatch of small sharks known as Pacific dogfish; in the past, Pacific dogfish would have been kept for food, their rough skins turned into a kind of sandpaper used to finish cedar carvings, their oil used to cure fishing longlines made of kelp. Seals and sea lions were once hunted widely and eaten with gusto—which also reduced seal and sea lion predation on salmon, herring, and eulachon—but now are eaten only rarely: their taste is too strong for most younger people. People from the communities tell me that this reduction in the spectrum of traditional foods reflects disruption wrought by racist federal policies enacted over the past 150 years—residential school, forced relocations, and so forth—which interrupted cultural continuity and acclimatized many people to the convenience and lower nutrition of store-bought foods.

Yet cultural diminution need not be permanent. People from the communities speak of certain practices as "muted" or "dormant"—not lost. Consistent with this view, the use of clam, root and berry gardens tapered off in the aftermath of colonization, yet the Hul'q'umi'num and WSÁNEĆ peoples of southern British Columbia have recently begun to restore clam gardens in their territories.

Other enhancement practices have always remained active. These include the transplant of culturally significant species—Dungeness crabs, herring (via eggs), and seaweed among them—to boost populations where needed. Until recently, the Nuxalk people enhanced spawning habitat for salmon and eulachon by clearing logs and other flood debris from rivers. Yet that practice ended when Fisheries and Oceans Canada (DFO)—the federal agency responsible for fishery management—threatened to charge them for "disturbing" those habitats. The Nuxalk see this forced restriction to their traditional practices as a contributor to stock declines. But not all that goes poorly remains that way. DFO is changing, or at least giving it a serious try, as a new generation takes over from the old guard. Perhaps it is only a matter of time before misguided federal policies are revoked and replaced with better ones that integrate time-tested principles of Indigenous management.

Critically, the hereditary chief system is very much alive. Much of my time in the Indigenous communities is spent working with chiefs who interpret traditional laws and, in collaboration with technical staff, integrate these laws into fishery management. The chiefs' long-term tracking of different resources is akin to taking the pulse of the ecosystem. When the pulse drops, the chiefs decree fishery closures within their territories, and most community members abide.

The cultures of many First Nations—their language, laws, and stories—have proven to be incredibly resilient. When people from these Nations refer to plants, animals, mountains, and water bodies as "all my relations," they mean exactly that. And they mean it just as much whether speaking in a boardroom, forest, canoe, or shopping mall. The fact that they drive big trucks, big boats, and big snow machines and fly on jets to Vegas only proves that all of us are part of the same modern world. Sure, that also means that both Indigenous and settler peoples have become contributors to global problems, but to leave it at that would trivialize the issue. The real point is that Indigenous peoples have begun to rebuild their political strength in Canada and other places, which opens the possibility for stories about kinship towards non-human species to become more universal than the collective psyche of industrial civilization has been willing to accept so far.

Traditional stories, of course, are not mine to retell. There are strict protocols about ownership of a story and the right to share it; in this book I show only glimpses approved by people to whom the stories belong. But what is mine to share are the insights I have gained as an ecologist who spends time with modern Indigenous people. Many insights are science-based, yet some are not—and that is a good thing, because it sure is going to take a lot more than science to get out of the global hole we have been digging for ourselves.

Individuals, of course, vary within cultures. As Cherokee-Greek writer Thomas King has pointed out, there are many instances in which Indigenous people have engaged in destructive industries and committed blunt violations of their traditional laws—sometimes because they were bamboozled by crooked political leaders, sometimes because they have been raised outside their own culture, and often because of a desperate and legitimate need to climb out of the poverty that fosters substance abuse and suicides. And, of course, there are plenty of individuals from settler cultures who are great environmental stewards. Yet a fundamental characteristic of Coastal First Nations and other Indigenous peoples is their explicit, collective, and culturally based ethos for kinship and responsibility towards the non-human world that goes back for many centuries. These are the ripples and powerful waves that emanate from a long line of ancestors speaking on behalf of Earth, Ocean, and Sky, with conviction and clarity.

I have been exposed to these views only because of the tremendous generosity of the First Nations I have come to know, who have allowed me to go out and get wet and dirty alongside them, who have shared their knowledge as we conduct collaborative research, or hunt and butcher ungulates, or gather medicinal plants and catch fish.

And, in the process, my Indigenous friends and colleagues have shared their own stories about being in a world that has been tough on them, but that appears to be on its way to becoming less racist and more inclusive.

These are stories about our origins.

About colonization and its destruction.

About healing and reconciliation.

About resilience in the modern world, where different cultures can, and should, intermingle.

And these stories have become part of a journey of my own. One that ultimately reshaped my view of where humanity stands and *could* go.

SMITH BROS

4 Reawakening

Waterfalls, forested slopes, and granite cliffs gather into ridgetops. Steep mountains shoot up from the estuary, fjord, and lake. Here, at the head of Rivers Inlet, the Waanukv River drains Oweekeno Lake. One hundred and seventy meters deep, on average, and nearly 100 square kilometers large, the massive lake is a critical spawning area for sockeye salmon along the Central Coast of British Columbia. People here have always shared the coast with grizzly bears, both species brought together by the Waanukv River to celebrate their shared connection to fish.

It is mid-October of 2018. As the boat docks at Wuikinuxv village, I take a moment to thank the surrounding landscape for having shown our daughter, Twyla Bella, that people, river, and ocean can be deeply interrelated with each other.

Here, 3½ years earlier, Twyla Bella, 11 years old at the time, joined our field research crew. She participated in deepwater surveys of rockfish using a towed video camera operated from a small boat and helped extract ear bones from specimens so that we could age individual rockfish. My wife, Gail, was also on that visit. The three of us joined Wuikinuxv people in a meal of eulachon (or "ooligan," as First Nations commonly refer to the species)—the small, fatty, silvery fish that spends most of its life at sea and spawns in rivers and that is a cultural lifeblood for many Indigenous peoples. For the first time in 20 years, the Waanukv River had provided enough eulachon to support a small harvest. But since that meal, the abundance of eulachon has subsided once again.

The Wuikinuxv say 10,000 people lived here before colonization, thriving on eulachon, salmon, and other riches from the sea, river, and lake. They also say that old-growth forest once blanketed the lower mountain slopes and valleys, an ecosystem that included gigantic Sitka spruces and red cedars, some of which were more than 500 years old.

Colonization changed this. The Wuikinuxv of today number only 80 people in the village and 300 more in towns and cities. Shrubs and trees along the river cover the remains of many who died during the late 1800s, killed by epidemics the colonists had brought. Sisulth (Frank Johnson), Chief Councillor and one of my hosts in the village, shares a telling image of that history during one of our conversations: "People would die while walking down the path and were buried right where they fell. That is why there are so many dips in the forest by the river, shallow graves that caved in."

As the Wuikinuxv suffered death and sickness, the landscape became vulnerable, unprotected by its human relatives. In the past century or so, industrial loggers have removed big swaths of forest that had held the ancient trees. Industrial fishers have depleted salmon runs. And, while fish returns in 2015 were encouraging, strong runs of eulachon have been absent for so long that parents worry that their children may never experience their own harvest and the traditional process of turning eulachon into grease: a most cherished food, medicine, and trade item.

"We used to get all the ooligans we needed," Kanilkas (Alvina Johnson), a practitioner of Wuikinuxv knowledge and Sisulth's wife, tells me. "We knew exactly what it would take to fill a bin for making ooligan grease. Enough for the winter. Enough for family. Then we would stop." After a pause, she adds wistfully, "But it's been a long time since we've been able to do that."

The changes are large, yet the people are reclaiming their agency.

Sisulth and Kanilkas keep their house heaters set on high—much too hot for those of us who grew up enjoying basic human rights. Yet both Sisulth and Kanilkas are residential school survivors. To them, the ability to be as warm as they please is a reminder that the worst of the horror has ended. That they will never be forced to return to the cold prison of their childhood.

Yes, the Wuikinuxv, the salmon, the Sitka spruces and red cedars, the bears, and even the eulachon are still here. The dozen or so children who attend the local school are taught their ancestral language, *W'uik'ala*, so that they may begin to recover the riches stolen from their grandparents. Carvers continue to gather wood from the forest and transform it into poles that capture Wuikinuxv history. Beside swaths of younger forest regenerating from industrial forestry, you can still go to the mouth of some streams and stare, in awe, at Sitka spruces and red cedars with trunks greater than two meters in diameter. Enough salmon are left to support bears and the remaining people, though just barely. And bears and people continue to live as relatives who must give each other space; to do their part, the people keep in their kitchens handheld radios that often transmit warnings about grizzlies walking down the street towards the river.

And every spring, Kanilkas walks down to the river to sing. To release into the atmosphere the *W'uik'ala* words and melodies that might merge with the mist of surrounding waterfalls and bring the eulachon back.

Year after year for the past decade and a half, the Nation's fishery technicians have used fine-meshed nets during late winter and spring to trap and examine the larvae of the fish that hatched in the lake or upper river and are drifting out to sea. The larvae include eulachon, which signals hope. Eulachon still spawn here every year, albeit in small numbers. People must prepare to welcome back their abundance, when the time comes.

The collapse of eulachon throughout the central and southern coasts of British Columbia is poorly understood. The species has tremendous cultural significance for First Nations but no market value. Consistent with Canada's history of institutionalized racism, DFO has a long track record of relegating concern, research, and recovery measures for eulachon to the lowest levels of priority. But we know that the decline correlates with a rise in industrial trawl fishing for shrimp, which captures eulachon as bycatch. Prior to 1996, that fishery was nearly lawless. Shrimp trawlers could operate year-round with no catch limits. The amount of eulachon and other non-target species they might have caught during that era was surely great, but it was undocumented. Rising sea surface temperatures have probably exacerbated eulachon declines.

My colleague Ts'xwiixw (Megan Moody) from the Nuxalk Nation, whose graduate work in fisheries examined the eulachon collapse, has described to me the origin of the problem as a "perfect storm" unintentionally created by DFO policies of the day for protecting commercial fisheries. Knowingly or not, such policies were built on the serial depletion model which, as unraveled by Ts'xwiixw's research, had unfolded this way: in the mid-1990s, salmon stocks were low and business was bad for commercial salmon fishers. Yet shrimp were abundant, and (in DFOs view) not enough fishers were targeting them. This made no sense (again, in DFOs view) because shrimp prices were high and fishers had unused trawling licences. To "remedy" the problem, in 1996 DFO opened to shrimp trawlers Queen Charlotte Sound, an offshore area off the Central Coast. But that was not enough. As Ts'xwiixw wrote, "DFO instigated a Pacific salmon license buy back in 1997. As a result many fishers began utilizing their shrimp licenses resulting in more shrimp licenses being issued." The unintended consequence: eulachon that use Queen Charlotte Sound, of which a majority appears to spawn in Central Coast Rivers, took a big hit. The bycatch of eulachon in Queen Charlotte Sound—90 tons in 1997 alone—broke a record for the entire coast. In Ts'xwiixw's words, "It is unfortunate that the largest bycatch occurred in the offshore areas inhabited by Central Coast eulachon, as they are some of the smaller eulachon populations." DFO closed Queen Charlotte Sound to shrimp trawling in 2000, which was a good thing. Yet other areas remain open to trawlers, and we know little about the extent to which Central Coast eulachon use those areas.

So I am back in the village of Wuikinuxv to speak with the people about how they might want to prepare for the day when eulachon return to the Waanukv River in harvestable numbers. Which traditions should people draw on to decide when and where to harvest, when to stop, whether to harvest at all? Should the first wave of fish, which are females, always be left unharvested so that they can lead others upstream? How might the Waanukv River become more inviting to the sacred fish?

The answers to these questions are in the cultural memory of the Wuikinuxv.

About a dozen people have gathered at the band office to discuss eulachon. Jennifer Walkus, former fisheries manager for Wuikinuxv and a person committed to integrating traditional wisdom with scientific ways of knowing, speaks about growing up in the village in the 1970s, when eulachon ran strong:

> When I was six I had my first job. When ooligans came up our river in massive numbers, my family would dip net, and I would have to separate the males from the females. The females are fattier, and we would turn those into grease. The males we would smoke.

Stephanie Henry looks out the window towards the river and says, barely suppressing a giggle, "We used to stand by the shore and dip net by lamplight through the night. Once I was seven months pregnant, and my family kept asking me to stop dip netting, warning that I might go into labor. But the power of the fish, the satisfaction of gathering our traditional food were so strong that I kept going."

Stephanie and Jennifer are roughly my age. They are among the youngest keepers of firsthand eulachon lore for Wuikinuxv. So the discussion turns to the need to interview elders and use their experience and traditional knowledge to develop a management plan to be used by the people when eulachon return.

The truth is that we do not know if and when large numbers of eulachon will come back to the Waanukv and other rivers—such as the Bella Coola River, which is home to the Nuxalk people—where their gifts have been deeply missed. Meanwhile, several rivers to the north and one river to south, within the territories of the Haisla, Nisga'a, and Dzawada̱'enux̱w Nations, have continued to provide eulachon. By trading seaweed, herring eggs, salmon, and other gifts with those Nations, the Wuikinuxv and Nuxalk have been able to maintain some connection to the sacred fish.

My hope, I tell those gathered, is that DFO will step up to its responsibilities to promote eulachon recovery. The federal agency has been reluctant to extend closures for shrimp trawlers beyond Queen Charlotte Sound, arguing that there is no evidence that eulachon caught in trawls originate from the Waanukv or Bella Coola rivers, rather than from rivers to the south and north that maintain strong runs.

Whatever happened to the precautionary principle?

Oh, yes, while eulachon are crucial to First Nations, they are unimportant to commercial markets. So far, uncertainty has been a convenient rationale for DFO managers to bias decisions in favor of those markets.

But then there is the tremendous capacity of DFO scientists. That includes geneticists who have been working hard to improve techniques for identifying the river of origin for eulachon caught as bycatch. Within the next couple of years, we may be able to estimate the extent to which bycaught eulachon come from diminished stocks, which should provide evidence for the need to further exclude trawl fisheries from specific areas. But the required resolution in the genetic data is not here yet. And people have waited far too long for meaningful action to promote eulachon recovery.

A huge source of frustration for Central Coast First Nations is that science already highlights a pragmatic way to reduce the bycatch problem—right now. Robert Hannah and other researchers with Oregon's Department of Fish and Wildlife, concerned with their state's own bycatch problem, ran field experiments in which they added LED lights to the lines that attach trawl nets to boats. The results were nothing short of amazing. The lights did not diminish the catches of shrimp, the species targeted by trawl fishers, yet they did alter the behavior of eulachon in ways that reduced their bycatch by an astounding 90%. As a bonus, bycatch also dropped by nearly 60% for juvenile rockfish and by 82% for adult darkblotched rockfish, an overfished species in which individuals can live more than 100 years.

Evidence for the tremendous conservation benefits of LED lights was so strong that no time was wasted after Hannah and colleagues published their work in 2014 and 2015. As the Oregon Department of Fish and Wildlife described in public communications (my emphasis),

> Upon the discovery of their effectiveness in bycatch reduction, the use of LEDs was immediately adopted by nearly all shrimp trawlers in Oregon. The use of LEDs is a leap forward in reduction of bycatch within the shrimp fishery. This was particularly timely as eulachon smelt, a historically common bycatch of pink shrimp trawling, was listed as "Threatened" under the Endangered Species Act (ESA) in 2010. *With the strong support of the industry, the use of LEDs became required for Oregon shrimpers beginning in 2018.*

And Hannah and colleagues wrote,

> As word of the results from our study spread through the California, Oregon, and Washington shrimp fleets, numerous vessel operators began buying and attaching green LED lights to the fishing lines of their trawls, reporting results very similar to our research findings.

Despite this scientific evidence, DFO has yet to make LEDs a requirement for shrimpers. According to Canadian fisheries regulations, lights cannot be attached to any commercial fishing gear, except that used to catch squid, presumably because they might attract some species and make them easier to overfish. Yet that concern clearly does not apply to eulachon and other species bycaught in shrimp trawls. This regulatory loophole may be keeping Wuikinuxv and Nuxalk people from reawakening their ancestral right to care for, and be cared for by, eulachon. Central Coast First Nations, of course, are strongly advocating for the implementation of LED lights on trawl gear, and—as of the spring of 2018—DFO managers have begun discussions about amending the regulations. Such legal processes can move at a discretionary pace. They can be fast or painfully slow. Whichever pace unfolds, it will reflect the extent to which Canada is committed to reconciliation with Indigenous peoples.

As the meeting begins to wrap up, people talk about how the Wuikinuxv once used canoes to carry sand back to river channels where strong currents had washed the original grains downstream. Sand is an essential substrate for eulachon to deposit their eggs. Its restoration into bare river channels ensured that the fish felt welcomed back. That practice, dormant for several generations, wants to awaken.

As I am about to exit the band office, I pause on the stairway before an old black-and-white photo of a house-front pole depicting Raven, a powerful spirit animal for Coastal First Nations. The caption reads: "Ke'tit Village, Oweekeno Lake, 1890."

No one lives in villages by Oweekeno Lake anymore. And I have walked along the somber river shore that Sisulth described.

Yet on the way to the band office today, I also walked past the longhouse where ceremonies and potlatches now take place. In front of the longhouse stands a new pole, raised only three months before. Carved in red cedar and at least three times taller than I, it re-creates the image in the historical photograph. Raven reawakened. Staring towards the waters that eulachon and salmon travel to their spawning grounds.

YAMAHA
90

5

The Exuberance of Herring

It is an early morning in March 2015, and I am in the company of two members of the Kitasoo/Xai'xais Nation: 71-year-old Charlie Mason (Neasmuutk Haimas) and his 10 year-old grandson, Dean Duncan. The temperature is 4°C. It is raining hard. Among tools and fishing gear, shelterless from the cold downpour, we are traveling in Charlie's aluminum punt powered by a 90-horsepower outboard along a channel of British Columbia's Central Coast. Our wake slaps against a rocky shoreline from which forested hills rise to become open mountaintops.

We had left Klemtu, Charlie and Dean's home village, about an hour earlier. At the end of a narrows where strong currents support a wealth of urchins, sea stars, kelps, and orange sea cucumbers, I recognize the beach where my companions' ancestors had carved animal and human-like shapes into intertidal boulders. As we zoom past, I recall the sinuous designs, set on black rock fringed by white barnacles and green bladderwrack seaweed. For me, who grew up in a faraway city and first experienced this coast as an apparent wilderness, the petroglyphs are a reminder that our surroundings are inseparable from human culture.

I keep my back to the wind while Charlie tells stories. This morning, perhaps because of Dean's captive presence, his stories revolve around the main activities of his youth—hunting mountain goats and deer, gathering eggs from seabird colonies on exposed rocky islets, harvesting other marine foods with his maternal grandfather, learning the ways of commercial fishing with his father.

That way of growing up, Charlie asserts, was a huge contrast to the comfortable distraction of electronic screens and gadgets that enrapture Dean's generation—which happens to be the same as my daughter's—so all of this is very relevant to me. "Everything is too easy for them. Turn on the switch and you got heat. I was always hand-sawing and carrying firewood, even when the shed was full!" Sitting on an old garden chair at the back of the punt, Dean smiles shyly, his teeth chattering in the cold rain, looking unsure about what to make of his grandfather's banter.

Charlie has the authority to speak of such things. As he was growing up, his own grandfather steeped him in traditional stories, protocols, and practice about resource management, and gave him the enormous gift of honor and responsibility. Charlie inherited his grandfather's hereditary chief name, Neasmuutk Haimas, meaning "provider" or "the person who gives" in the Xai'xais language. Ever since receiving it, Charlie has embodied the spirit of that name, serving as one of the main harvesters of traditional foods for his community.

His family history has made Charlie a bridge between eras. Before colonization forced people from different villages to amalgamate into centralized communities like Klemtu—where two sets of related peoples, the Kitasoo and the Xai'xais, came to live together in 1875—Charlie's maternal grandparents lived in Q'enxv, a village tucked inside Kynoch Inlet. Kynoch is a spectacularly steep fjord where winter snows push mountain goats to lower elevations. One of Charlie's grandfather's roles was to sit hidden along a narrow path between cliffs, waiting for mountain goats to pass, then kick them off the cliff towards other hunters below. Anybody able to sustain such patient immobility in the cold of winter while remaining ready to kick hard at a sudden and precise moment must have been a remarkable mentor, and Charlie remains committed to passing on to younger people what he learned from his grandfather.

Charlie describes his position among his people this way: "You are a chief, not a king." And he means it. Over the years I have known him, I have admired his constant generosity and inclusivity. Without prompting, he frequently offers his knowledge and experience to those willing to listen, regardless of whether they are Kitasoo/Xai'xais or from a different culture. Charlie recognizes that the stories he carries can connect traditional teachings and the modern, rapidly changing world. If stories are not passed on, the death of a storyteller can bring the demise of irreplaceable knowledge. Fortunately, leaders from the four First Nations with whom I work recognize this risk. They have been preserving the stories and wisdom of their people, archiving into computer databases the transcripts of historical and recent interviews with elders and other keepers of traditional knowledge.

Keeping your boat from sinking and your net from being raised empty requires a detailed understanding of your fishing grounds. A few years back, when interviewed for the Kitasoo/Xai'xais archives, Charlie described how he had gained that understanding in the traditional way:

> It was quite interesting, from the time I was probably about twelve, thirteen years old. My Grandfather [and I], we'd travel from place to place, looking for fish. And all of us, being young, we'd read comic books and lay down, go to sleep. Grandfather came down, right down to where I was laying down. "Come up top." After I sat down. "You stay right there. One of these days you are going to own your own boat, or you are going to run a boat. How are you going to know where to go, how are you going to know where your land is, where the rock piles are? How do you know where the safe places to go? How do know where you are going to anchor out of the storm? How do you know?"
>
> Because there was nothing like this in those days...no radar, no sounder, no charts. All landmarks. You see this area offshore? When you get to it you line up this island with this

> mountain, just straight down the middle, you start setting your gear out this way. You set it side by side. Another way to go further out, you line up that mountain with that one there and just sort of change the landmark. Quite interesting, they used to have [exact knowledge] without even using the sounder, you used to be able to go set your gear right there and you remember the places.

I have been sea kayaking along the coast of British Columbia since the early 1980s, which gives me some appreciation for what it takes to stay on route and anticipate—in the fog and without electronic instrumentation—where submerged rocks might suddenly cause a passing swell to explode. Yet the kind of seascape knowledge that Charlie holds belongs in an entirely different league: a realm attainable only by those raised within a culture that cultivates, from an early age and with strict discipline, the power of detailed observation.

Shortly after passing the petroglyphs, Dean, Charlie, and I arrive in Kitasu Bay, where our day will center around Pacific herring: a silvery, fatty fish found throughout the north Pacific Ocean. When not overfished, Pacific herring can live nearly two decades and exceed 30 centimeters in length. Yet commercial fishers appear to have left their signature. Our ongoing studies of spawning aggregations in areas used by the Kitasoo/Xai'xais and Heiltsuk Nations suggest, so far, that the largest herring are only 23 centimeters long and the oldest individuals only 11 years of age. Pacific herring spend most of the year in mobile schools that are difficult to pinpoint. But all that changes during spring when (unless stocks have been overexploited) the fish predictably aggregate to spawn along shorelines, mainly at shallow depths—which makes herring relatively easy to harvest that time of year.

Archaeological data indicate that Coastal First Nations began fishing for Pacific herring at least 10,000 years ago, and the species has been a staple for at least 2,500 years throughout British Columbia and adjacent areas. The accumulation of bones in middens suggests that herring remained steadily abundant throughout that period. Herring numbers did fluctuate over time, likely in response to cyclical variation in climate and ocean conditions, such as those caused by oceanographic processes like El Niño–Southern Oscillation. But that variation appears to have remained relatively small, such that even during the low points in the cycle there was plenty of herring for First Nations to harvest. That steady abundance of herring was maintained over time largely because of intentional practices for resource stewardship enacted by hereditary chiefs. As they did with other species, the chiefs used their extensive knowledge and traditional laws to determine where and when herring fisheries may or may not take place, as well as the amount of harvest.

Since colonization, herring populations have undergone much wider fluctuations and huge drops in average abundance over time. Industrial fisheries have contributed to that trend by exploiting adults that are about to spawn and that have crowded into very small spaces where they are easy to fish. Those spaces—generally a few bays or channels—may sometimes contain most of the remaining fish in an entire region. The result: commercial fishers can sustain large catches and keep making money while depleting the stock, until a sudden crash occurs. As a result, herring declines have become steeper, stronger, and more frequent than in the past.

When herring spawn, females release eggs that attach to kelp, smaller seaweeds, or rocks while males release clouds of milt that fertilize the eggs. To enhance herring productivity, Indigenous fishers add novel substrates—such as conifer boughs or transplanted kelps—to which herring eggs can attach over sandy bottoms, where they otherwise would not spawn. Coastal First Nations have long understood that very few eggs survive to become adults and that, consequently, egg harvests remove only a small proportion of individuals likely to reproduce in the future. For these reasons, both adult fish and herring eggs have been historically important in traditional diets, yet only herring eggs remain the focus of harvests by modern First Nations.

As Charlie, Dean, and I travel west into Kitasu Bay, Marvin Island, which is known to the Kitasoo/Xai'xais as Láktiha, comes into view towards the bay's south end. The sight brings back memories of when I began dive surveys for rockfish in the area two years earlier. During a dive off Láktiha's north shore, I was working with Ernie Mason, who is Charlie's son and Dean's uncle, while Ernie's wife, Sandie Hankewich, drove the boat with one hand and cradled their one-year-old daughter, Jesse, with the other. Suddenly, a massive herring school darted through the clear water. I was just beginning to wonder what might have spooked the herring when a blue shark, perhaps two meters long, appeared in the distance. It is rare for me to see sharks underwater, so I took in the sight for as long as I could, reveling in the vibrant tension that exists between top predators and prey.

For centuries, Láktiha was a seasonal camp for the gathering of herring eggs. When he was interviewed for the Kitasoo/Xai'xais archives, hereditary chief Percy Starr, whose advanced age now prevents him from harvesting resources in the field, spoke about his youth, when up to 30 families lived at Láktiha for five to six weeks at the time during herring egg-harvest season. With the advent of faster and better boat engines, residents of Klemtu now tend to visit Kitasu Bay as a day trip. Yet, to this day, the area around Láktiha continues to produce most of the herring eggs that the Kitasu/Xai'xais use as traditional foods, for commercial purposes, and to trade with neighboring First Nations for traditional foods that are locally less available, such as the grease of eulachon. And the island remains a seasonal camp for the cultural education of young people under the tutelage of elders like Charlie.

Defending this cultural and biological lifeblood has not been easy. In British Columbia, industrial fisheries for herring go back to 1877, and the years between 1954 and 1967 stand out for the degree to which the fisheries devastated herring. Retrospective estimates suggest that, during that period, nearly two thirds of herring biomass were removed each year, triggering catastrophic declines among most populations.

Let's pause and allow that to sink in. Nearly two thirds of the herring biomass removed, year after year, for 13 years in a row. And all those fish were caught to be processed into fish meal and oil, meaning that the overexploitation of an ecologically significant species that is also a cultural keystone occurred not to feed people directly but, rather, to produce fertilizers and animal feed—key ingredients for industrial agricultural practices that exacerbate climate change and other global problems.

As someone who believes that humans are not innately destructive—but who understands that, under certain circumstances, we are easily swayed into being cognitively immature and misguided—I do wonder why such blatant damage occurs. My guess is that lack of information played a role in herring overexploitation. Fish in the ocean are notoriously difficult to study, especially when it comes to estimating their actual abundance. As fishery scientist John Shepherd famously quipped, "Counting fish is like counting trees—except they are invisible and they keep moving." And methods for estimating fish numbers in the 1950s and 1960s were more rudimentary than they are now. Still, if you are unsure of how many fish are out there and how many you are leaving behind, why not err on the side of caution and slow down?

Poor scientific data are not the essence of the problem. Market impunity has the capacity to overtake awareness of and personal connections to local places and ecosystems, and to undermine Indigenous ways of being in the world. The colonial approach to resource management has a long history of dismissing the well-being of communities like Klemtu. Simply put, the declines of herring and other culturally significant species are rooted in racism. The corollary of this gloomy statement is incredibly simple—and optimistic. The greater the degree to which Indigenous people become legitimate governance partners in the modern management of fish and other resources that derive from their ancestral territories, the more likely the sustainability of that resource.

By 1967, the decline in herring became so glaring that even DFO could no longer pretend otherwise. So it called for a "cease fire," ordering the industrial fleet to halt its fishery. The reprieve lasted only four years. When industrial fisheries resumed in 1972, they shifted their focus from catching fish to be processed into fish meal and oil to targeting herring to extract their roe, which at least meant that the catch would directly feed people. Importantly, DFO reduced the allowable exploitation rate to one fifth of the estimated biomass. Though certainly an improvement, this new management approach was not conservative enough to promote recovery. And after 1972 the industrial fleet expanded its area of operations, fishing waters it had not previously exploited, including Kitasu Bay.

Under this new regime, industrial fishers exploited Kitasu Bay every year from 1972 until 1990. Since 1990, the Kitasoo/Xai'xais have applied their traditional laws, civil disobedience, and other legal tools to deter the industrial fleet from Kitasu Bay, successfully doing so for 23 of the last 29 years. Herring abundance in Kitasu Bay remains, on average, below historical levels. But it is probably because of these Indigenous efforts that Kitasu Bay currently is the strongest spawning area for Pacific herring in the Central Coast of British Columbia.

As Charlie, Dean, and I proceed further into Kitasu Bay, the ecosystem role of herring becomes spectacularly clear. Bald eagles perch on shoreline trees, and from our punt their white heads look like a nearly continuous dotted line against the dark green canopy. Ravens and crows saunter along the beach, where low tide has exposed rocks covered in herring eggs. Clouds of gulls and large groups of Steller sea lions commingle in the small coves where amassing herring roil the water. Farther out in the bay, three humpback whales dive and surface and dive again. Flocks of seabirds—black and white surf scoters, their males with bright orange beaks, and common murres in drab gray winter plumage—float as far as the eye can see on a gray surface the rain pricks into tiny ephemeral craters. The scoters, crows, and ravens are here to eat the herring eggs, while all the other animals and birds are here to eat the herring themselves. A feast that allows a plethora of predators to refuel at winter's end.

And that is only what we can see above the surface. Underwater, crabs, snails, and sculpins, and other small predators, no doubt, are feasting on herring eggs, while big fish like Pacific halibut likely are snatching any adults that drop down within their depth range. Charlie, of course, knows that herring egg season is a prime time to harvest herring predators. Halibut can be fished most efficiently this time of year, when they move into shallower, nearshore areas to feed on spawning herring.

Charlie glides the punt against the gently sloping rocky shoreline and encircles a herring school in a seine net that we haul in almost immediately. Within moments, the punt's floor is covered with wriggling bodies. Next, we use these herring to bait about 100 hooks, which we attach with terminal rigs called gagnions to a half-kilometer-long groundline. Charlie lines up his landmarks—a specific point at the base of a prominent hill to the south and another point out to the northwest—and we drop one end of the groundline, which is attached to a weight and marked by a second line connected to a surface float. From there, we slowly head towards the second landmark while paying out the remainder of the groundline at a depth of about 70 meters over sandy bottoms where halibut are likely to be caught, and over isolated rock piles that rockfishes use.

While the groundline soaks, we beach the punt and step ashore. We spend a few hours collecting beach boulders, chainsawing boughs from hemlock trees in the forest, and collecting blades from two species of kelp—feather boa, which the Kitasoo/Xai'xais call *luggi*, and giant kelp, *cluquala*. We then return to the punt and motor to a small bay near Láktiha where Charlie expects herring to spawn over the next day or so. We tie the kelp blades and hemlock boughs to a line, much like we placed hooks for halibut on a groundline.

Then we wrap the boulders in pieces of old fishing net to turn them into anchors, and set the line on the bottom, five meters below the low-tide line.

This simple action—the sinking of kelp blades and hemlock boughs attached to a groundline—encapsulates centuries of seasonal rounds by Coastal First Nations. Charlie's ancestors would have used dugout canoes instead of a motorized punt, hooks made of hardwood or bone rather than metal, and lines made from the cured stipes of kelp rather than nylon and cotton. Yet the fundamental practice remains the same.

As herring spawn over the next day or two, the eggs will attach to blades and boughs, each type of substrate infusing a unique flavor to the eggs. Charlie will let multiple egg layers accumulate over the course of a few days, and then pull out the line to take home the bounty. Along the way back and at the home dock, he will toss some of the eggs back into the water, hoping to expand the number of places where herring spawn, perhaps helping to revitalize some of the runs that industrial fisheries depleted.

On our way back to Klemtu, we head for the halibut groundline, grab the float and start pulling. We have no mechanical hauler and, after letting me do what he deems to be my share of the grunt work, Charlie motions me to step aside and takes over. I almost blurt out something about letting my younger body do more of the work but then I recall his self-description: "You are a chief, not a king." I'm good with that. This way I get to enjoy the thrill of gaffing fish after fish, off the line as they surface.

Our catch is 53 halibut, several weighing 30 kilos or more, 14 quillback rockfish, and a Pacific cod. A much better haul than when I longlined with Charlie in the same area the previous summer—that is, in the absence of herring—when the catch per the same number of hooks ranged from only one to seven halibut, plus a few rockfishes.

As we rearrange the gear on the boat, Charlie tells me that an uncle taught him to make an infusion from the bark of devil's club—a prickly yet incredibly medicinal and nutritional forest plant considered to have special powers—and soak his groundline in preparation for fishing. The last time he did so, Charlie tells me, he caught 122 halibut with 200 hooks. Despite that huge success rate, he does not practice the devil's club soak very often. In fact, I get the sense he does so rarely, if ever, these days. When I ask why, Charlie responds: "You get way more than what you need."

The rain is relentless. It does not let up as we arrive at the village's main dock. As we were heading back from Kitasu Bay, Charlie radioed our approach over the VHF radio, and community members already are waiting for us to distribute the halibut catch. I had better be quick in measuring the length of all the fish and weigh as many halibut as possible, ensuring I collect the necessary data for our ongoing research. Already, Charlie's wife Edna is teasing me to hurry up with the "science stuff" so people can get out of the rain. Fifty-three halibut plus a whack of rockfish. Good luck with that. So we negotiate: I will measure and record enough halibut for the elders to take home right away, and the remaining folk will return later for their catch. Which leaves me alone in the rain and the gathering dusk, extracting ear bones and tissues from rockfish, measuring halibut after halibut, covered in fish slime yet delighted with the riches of the day.

Herring season is far from over. Charlie and crew will make more trips to Kitasu Bay to set and pick up more lines, and I will tag along. And when that is done, the herring predators will still be around in large numbers. Coastal First Nations say sudden noises or blood in the water disrupts spawning, so they adhere to strict protocols to prevent such disturbances. But when the spawning is finished Charlie will return to hunt the lingering scoters, seals, and sea lions.

The following March, 45 kilometers east of Kitasu Bay, I am at the end of a fjord, floating nearly 30 meters below the surface, midwater off a submerged cliff that resembles the terminus of a tidal glacier—white, absolutely white, covered by millions upon millions of Pacific herring eggs the fish deposited over the preceding days. The herring adults have moved on, leaving the underwater space quiet. Peaceful. I swim away from the wall to take in a wider view through the clear water. One of my colleagues, Markus Thompson, a graduate student from Simon Fraser University, becomes a tiny silhouette engaged in staccato movements as he propels himself with short bursts of fin kicks and pauses, intermittently, to collect eggs for later analysis under a microscope. These sorts of dives, due to their depth, are short. Yet their imprint in memory lasts a lifetime. We surface and gather under the splash of a waterfall bouncing off the mountainside and into the sea. Our cries of excitement for what we have seen are loud and childlike.

What we saw during that dive was a unique beauty that none of us had witnessed before, and over the following days our team—which includes ecologist Dan Okamoto from Simon Fraser University and field biologist Tristan Blaine from the Central Coast Indigenous Resource Alliance—continues to document other stunning "herring walls." Yet, beautiful as they are, the herring eggs attached to steep granite and reaching as one continuous blanket into deep darkness might be a cautionary signal of ecosystem change, and fishers from the Heiltsuk Nation have asked us to investigate.

Herring typically deposit their eggs within five meters of the low-tide mark, but the walls we are studying hold eggs that span from the surface to depths of 40 meters or more. Such depth for herring egg deposition is very unusual, and its apparent increase in frequency was first observed in 2014: indirectly, by Heiltsuk fishers pulling up their egg-covered lines and anchors, and directly, by divers working with DFO. In response, Mike Reid, the aquatics manager for the Heiltsuk Nation, raised two questions to be addressed by Markus's graduate work: Does spawning deeper diminish egg survival? and Why have herring changed their spawning behavior?

I have worked closely with Mike for several years. A former commercial fisher and lifetime harvester of traditional foods, he possesses detailed knowledge of the coast. Some of his expertise he inherited from his father, Fred Reid, whose depth of local and traditional knowledge is legendary. I have sat beside Mike during many meetings with DFO that concern tough issues affecting Indigenous fisheries. Whenever such discussions start to become contentious, Mike's gentle eloquence invariably diffuses the tension, reminding everyone of our shared objectives and getting us back on task. And I often seek out Mike to double-check my ecological ideas, making sure that some beautiful hypothesis in my head is not luring me away from the real natural history he knows so well.

Mike's questions about unusual spawning behavior are embedded in a broader concern for herring populations. The primary areas where Heiltsuk fishers collect herring eggs—both as traditional food and to sell commercially—were hit hard by industrial fisheries for adult herring up until 2006 and, after a brief pause, again in 2014. Some Heiltsuk fishers have observed that this overexploitation has removed most of the older herring, reducing the population to primarily younger fish—which is consistent with age and size data that we have collected—and believe that the younger fish may be making some poor choices when selecting spawning locations. Also, likely in response to overexploitation by industrial fishers, British Columbia's Central Coast has lost half the spawning areas known to be active until the 1940s. That loss has been unequal across space. Very few spawning areas remain on outer islands, and most herring now aggregate to spawn in inshore bays and fjords. This means that a high proportion of the remaining members of declining populations have been crowding into fewer habitat types, which likely erodes resilience to industrial fisheries and climate change.

In search of answers to Mike's questions, we continue to survey the range of depths and other conditions in which herring spawn. Working with Ernie Mason and Sandie Hankewich, we also run a field experiment, diving to transplant egg-covered kelp blades to different depths; some blades are encased in small cages to exclude predators like crabs and small fishes. Over the following two weeks we collect these eggs at different stages of development.

Later, back at the laboratory, Markus studies the herring egg collections and pieces together the story told by the data. At experimental sites with predator exclusion cages, eggs near the surface are almost twice as likely to survive to full development than eggs transplanted to depths of 30 meters. Outside experimental sites, eggs collected from deep sections of herring walls are unfertilized and have poorer survival and fewer egg layers than those collected at shallow depths. The combined evidence gives a clear answer to Mike's first question: Yes, spawning deeper does reduce egg survival and hinders reproduction.

Answering *why* maximum spawning depths have increased in certain locations is much more difficult. During field surveys we had estimated rates of boat traffic that might disturb spawning aggregations and the densities of predators that eat herring adults, such as sea lions and humpback whales, and of predators that eat herring eggs, such as scoters. Yet our analyses conclude that neither boat disturbance nor high risk from predators explain why herring spawned deep at the locations we have examined. Deep spawn occurred only where two conditions converged: fjord walls with continuous rocky habitat extending from the surface to deep depths, and extremely dense spawning aggregations—over 100 females per square meter (as estimated from egg densities, which in some cases exceeded two million eggs per square meter). Having only one or the other condition did not lead to deep spawn. For instance, at places with high herring densities but no fjord walls, like Kitasu Bay, the herring had spread out horizontally along shallow depths, rather than towards deeper areas with isolated rock patches. Similarly, wherever there were fjord walls but few herring the spawn had occurred only near the surface.

These results hint at a potential connection between industrial fisheries and spawning behavior. The decline in spawning sites on the outer coast, which Heiltsuk fishers have observed over time, might be promoting crowding into the fjords and, indirectly, enhancing the chances that herring might spawn deep. But that still does not answer *why* herring spawn deep, especially if deep spawning appears to bring no reproductive benefits. So we are left wondering whether herring in fjords have faced some novel condition to which they have yet to adapt and that we do not yet understand.

Ocean warming is a candidate for that novel condition. So far, most observations of deep spawn by Heiltsuk fishers and research divers (including us) have occurred between 2014 and 2016, when an extreme slowdown of circulation in the ocean and the atmosphere caused The Blob—a 90-meter-deep mass of warm water covering 1,600 square kilometers—to stall in the northeast Pacific Ocean. During 2015 and 2016, when deep spawn appears to have been most prevalent, sea surface temperatures in British Columbia's Central Coast at the time of herring spawn (late March to early April) averaged 10°C, nearly 3°C warmer than during the previous 60 years. Since 2017, The Blob has dissipated and ocean temperatures, though still warming in response to climate change, have been cooler, closer to the average of the past six decades. None of the Heiltsuk fishers with whom I have spoken have reported deep spawn since. One can view The Blob as a kind of time travel, a glimpse into what future warming, combined with the ongoing loss of spawning sites in outer shores, might do to the reproductive success of herring.

One early morning during the 2015 herring season, I meet Charlie at the dock ahead of the crew of Klemtu residents, ranging in age from mid-20s to mid-50s, who is to join us that day. While we wait, Charlie begins to speak of how he will be teaching "the young guys" places to set lines for herring eggs. Suddenly, he stares out into the bay and recalls how in 2013, a very hot day at the end of summer abruptly ended an abundance in this very bay of pre-spawning coho salmon. That event was unprecedented in Charlie's experience, and he took it as a dire warning. "I am not a biologist, but I have been watching. I know."

There is no question that climate change will be challenging for First Nations committed to their traditional diets, and that Indigenous peoples on their own cannot solve the problems that originate with industrialized civilization. But maybe nobody is on their own anymore. Maybe the coalescing of different ways of knowing is beginning to take on genuine traction, possibly giving all of us a chance to create a more benign Anthropocene.

In the face of climate change and declining stocks, it is logical to manage fisheries much more conservatively and to follow the advice of Indigenous peoples who have studied and used those stocks for centuries. DFO has a long history of doing the opposite, although things might have just begun to change after a last set of kerfuffles.

During the 2015 herring season, as I was spending time in the field with Charlie, DFO opened Heiltsuk territory to industrial fisheries. To protest the opening, a group of Heiltsuk youth paddling a *glẃa*, or traditional canoe, crossed the waters from their village in Bella Bella to the local DFO office at Denny Island. A delegation of about 50 Heiltsuk community members, some dressed in traditional regalia and adorned with ceremonial war paint on their faces, followed in powerboats. The delegation included elected Chief Councillor Marilyn Slett and Stewardship Director Kelly Brown, who proceeded to occupy the DFO office with the intention to hold their ground until federal higher-ups came to their senses and agreed to close the industrial fishery. Slett and Brown spent two nights in the DFO office before tough negotiations with federal directors and ministers led to the closing of the fishery.

At the same time, DFO announced their intention to open Kitasu Bay to industrial fishers. Day after day, elected Chief Councillor Doug Neasloss and his technical staff attempted to negotiate with DFO over conference calls, but these conversations went nowhere. So the Kitasu/Xai'xais told DFO that they would physically block commercial fishers from setting their nets. Fortunately, DFO chose to avoid a confrontation on the water and withdrew their intention to send fishers to Kitasu Bay. People in Klemtu and Bella Bella could breathe a sigh of relief, happy that further conflict and court cases had been narrowly averted.

In recent years, First Nations have often won legal battles against the federal government over resource management. That is one way to assert Indigenous rights. Yet Brenda Gaertner, a lawyer for Central Coast First Nations, has remarked that "courts set standards, but reconciliation comes through agreements." In other words, the best fight is the one that is avoided and transformed, preemptively, into common ground. Consistent with that view, Doug Neasloss, Kelly Brown, and other people from First Nations communities speak of civil disobedience, either intended or realized, as difficult and scary: something they would rather not engage in but that they would certainly pursue again in defense of herring or other cultural keystones, should the need arise. So far, since the last herring standoff that has been unnecessary.

After the tension of 2015 dissipated, DFO scientists, who are incredibly skilled at what they do when unimpeded by the old bureaucratic guard, created a technical working group with the Heiltsuk Nation to develop a joint plan for herring management on the Central Coast. (My expectation is that the working group will soon expand to include the Kitasoo/Xai'xais, Nuxalk, and Wuikinuxv Nations.) As the 2018 herring season approached, that joint plan had yet to be completed. DFO might have been tempted to revert to calling the shots on their own but, much to their credit, they did not. Instead, they preemptively announced that there would be no fishery in the Central Coast, sending the following letter to "Chiefs, Councillors, Fisheries Managers, Technicians, Guardians, Biologists, and Harvesters" with a vested interest in the area:

> In the Central Coast stock assessment area...DFO and Heiltsuk were unable to reach a shared understanding of stock health or the potential to open both spawn-on-kelp and roe herring fisheries. In order to support the government's reconciliation agenda, including supporting co-management

> of fishery resources, taking into account the Heiltsuk's traditional knowledge of the ecosystem, and given the Department's risk of being unable to ensure orderly and well managed fisheries, DFO has agreed to suspend the roe herring fishery in the Central Coast for 2018. DFO will continue to collaboratively engage Heiltsuk in decisions about the Central Coast herring fishery. This collaborative work will be inclusive of opportunities for other sectors to access the resource when there is sufficient abundance.

After more than 150 years of institutionalized racism, to which DFO has significantly contributed, the positive significance of that letter cannot be overstated. And perhaps the change is here to stay. When the 2019 spawning season came along, DFO again respected the position of First Nations and did not open a commercial roe fishery in the Central Coast.

Sometimes I find myself on a boat with the engine turned off. The surface bubbles as herring aggregate; the sea turns light turquoise as males release clouds of milt. Steller sea lions and Pacific white-sided dolphins charge into the massive fish schools. Surf scoters rest nearby on the surface, clustered tightly by the thousands, waiting for the tide to drop so that they can feed on herring eggs deposited on intertidal rocks. Humpback whales breach all around us.

At times like this I stop thinking about the impacts of colonization, stop imagining the carbon dioxide being absorbed into the water. Temporary insanity, willful myopia—call it whatever you wish. All I know is that such moments help shape the stories I continue to tell myself about where humanity could go.

There are many definitions of joy. One that I particularly like is encapsulated each time food fishers from First Nations communities arrive at the docks of their home villages, where people, alerted by a VHF radio call, have gathered to receive the gifts of herring eggs and halibut. This is when the food fishers, the people who give, transform their hard work into cultural empowerment: a sign that, despite centuries of colonial efforts to annihilate them, Indigenous peoples are here to stay. And so full of humor, that most powerful source of human resilience! The way Edna hounds me to get on with

the halibut measurements. The way people at the dock load themselves with herring abundance and herring giddiness.

After all, to eat herring eggs at a potlatch, with a side of seal meat and scoter stew, is to take in a whole ecosystem into your body.

6

Sculpted by River and Story

The river had a violent birth. It burst forth beneath the Tchaikazan glacier, ruptured out of an ice cave sculpted by the sheer force of rushing water. Submerged boulders rumbled inside the current.

The river was a storyteller. It told of how Clark's nutcrackers got their wings, how mountain goats got their snowy fleece. It spoke of the hunter who discovered how to chip stone into projectile points.

The river gushed away from the glacier where ice worms and a thin layer of red algae thrived, ephemeral, in the relative warmth of summer. It kept its speed past an eroding gravel bar where flowers—paintbrush, mountain avens, fireweed—exploded out of grayness into bright red, pale yellow, vibrant purple. The water took with it chunks of that gravel, clumps of those flowers.

The precariousness of existence. The insistence of exuberance.

The river accelerated in the steeper gradients. It did not slow down until it reached the wetlands covered with the white orchid *Platanthera*. It continued through the forest where grizzly bears had left rubbing marks on the trunks of lodgepole pines and white spruces.

My brother Leonardo, our friend Brad, and I had followed this river on the east side of British Columbia's Coast Mountains, throughout its valley and on to its highest source, the summit of Mount Monmouth. We had come in search of a projectile point chipped from stone 10,000 years before. Carrying all of our supplies on our backs, we had been travelling with a minimum of food and no tent. This had allowed a welcomed closeness with the mountains and one another. We had waited out storms behind boulders, drinking tea and falling asleep on one another's shoulders.

The year was 1988. A time early in my adulthood and long before I connected with the people who have lived in this landscape for so many generations. Years before I even heard of human-caused climate change, ocean acidification, and the microplastics that have infiltrated just about every living cell in the ocean. A time of innocence, perhaps, when the course of the river was the only path I could understand. The river meant flow and transfer of energy and synergy and spiraling currents and all that runs through my veins.

Struggling for a last glimmer of youthful insight towards the end of the trek, I wrote this in my journal:

> Sitting by the glacier snout, it is easy to forget that the planet Earth is at the most desperate stage of environmental deterioration it has ever known. But all I have to do is remember the many British Columbia valleys with roads into them to be bluntly reminded by clear-cut logging scars of what state most of our planet is really in. We no longer have to pursue a single animal for days before killing it with our stone tools. We now have machines capable of killing whole ecosystems in a day. We are in a hurry to get who-knows-where. Our tools have evolved faster than we have. How many more valleys will have their ecological integrity disrupted by our loss of control over our own technology? Is it only a matter of time before the Tchaikazan becomes one of them? The din of environmental deterioration makes humankind deaf to what a healthy planet really looks, feels, smells, and sounds like. It obscures the projectile point that not long ago gave direction to our evolutionary path. At least for the moment, the damage is not fully done. The Tchaikazan River is still one of those chunks of Earth that have yet to give up.

Call it midlife crisis, scientific literacy, or a little bit of both. The truth is, three decades later I feel rather nostalgic for that "most desperate stage of environmental deterioration" of which I wrote at the age of 24. Sure, the logging was terrible. It ripped apart all sorts of biodiversity. It destroyed ecological processes that remove greenhouse gases from our atmosphere and store them into wood, leaves, and soil. Still, back then, when I perceived the advancing clear-cuts to be the pinnacle of environmental destruction, we still lived in atmospheric paradise—the tail end of it.

As I navel-gazed by the Tchaikazan River in 1988, the atmospheric concentration of carbon dioxide (CO_2)—the most abundant greenhouse gas produced by our fossil-fuel economy and the major driver of climate change and ocean acidification—was a mere 352 parts per million (ppm). Granted, a lot more than the 280 ppm of the mid-1700s, just before the Industrial Revolution, yet something we could work with. We blew past 400 ppm in September of 2016.

That was a brutal threshold.

Paleoclimatologists measure ancient CO_2 concentrations from air bubbles trapped deep inside the Antarctic and Greenland icecaps, infer prehistoric temperatures from fossilized plant communities, and map past sea levels from eroded landforms. Thanks to their research, we know that the last time the atmosphere held 400 ppm of CO_2 was 15 to 20 million years ago. The Earth of back then was 3–6°C warmer than today and had no polar ice caps. Its sea levels were so much higher than they are now—by the equivalent of an 8- to 13-story building—that none of today's coastal cities could have existed.

Sure, if you look out the window it is obvious that things are not nearly that dramatic, yet. Lucky for us, the Earth's response to the buildup of greenhouse gases is not instant, granting us some time to get our act together so that we may change course. But the same laws of physics that give us this period of grace also dictate that many of the choices we already have made, historically and up to this very moment, can haunt us far into the distant future.

The cumulative effect of the multiple and seemingly inconsequential decisions that individuals all over the world make during a single day—driving a car, riding a jet plane, lighting a propane stove, buying a new cell phone—will affect the next 20 to 50 generations. The reason is that two thirds of the CO_2 that we emit on any given day will spend decades to centuries warming up the atmosphere before the carbon cycle captures it back into plants, soils, and the ocean, while the remaining one third stays in the atmosphere for a whole 1,000 years. Also, the tremendous volume of water in the ocean requires centuries to absorb any warming in the atmosphere and, once warmed up, spends centuries releasing that heat back into the atmosphere.

So I try to let go of what was, to think of where we stand now, of where we can go from here.

The Bella Coola River has turned dark brown, and Jason Moody is stunned. "How did that happen so quickly? The water was a light and bluish gray when I set my gill net yesterday." I listen carefully, silently. I have known Jason—the son of the late Chief Qwatsinas (Spirit of the Raven)—for about five years. Over time, I have learned that his generous disposition and easy-going demeanor are conduits for deep knowledge of the land and culture of the Nuxalk Nation.

We are standing on the riverbank some 40 kilometers upstream of the estuary where the river drains into a steep fjord of British Columbia's Central Coast. By that estuary is the town of Bella Coola, where both settlers and Nuxalkmc live. Our location is just inland from the territories of the Tŝilhqot'in Nations, which include the Tchaikazan River. Both regions are culturally intertwined. The Nuxalk and Tŝilhqot'in have a long history of trade, in which the Bella Coola River is a major route between the coast and the interior.

It is early July, when the numbers of sockeye, spring, and chum salmon returning from the ocean to spawn in the Bella Coola River or its tributaries are peaking. The chums are still downstream from us, much closer to the estuary. Jason has picked this upper section of river to target sockeyes and springs.

We place Jason's aluminum rowboat at the edge of the water and prepare to check his net—if we can find it. Jason expects his gear to be gone, swept downstream by the torrent caused by the overnight two-meter rise of the river—the aftermath of a heavy outburst of rain that triggered an upstream landslide and muddied the water. Extreme events have occurred here before, Jason tells me, but now they are so much more frequent and unpredictable, like this quick rise of the river.

I almost blurt out something about how consistent this is with climate science—how studies X, Y, and Z have corroborated that rising global temperature have led to a greater frequency of floods, droughts, hot and cold spells. But I catch myself. Jason already knows.

Much to our surprise, things turn out to be alright—this time. We launch the boat and, within seconds, see floats bobbing in and out of the chocolate-colored water. The current has pushed Jason's net closer to shore, and some floats have sunk, but the gear remains mostly in place, anchored to a tree. I watch Jason maneuver the rowboat while he simultaneously checks his net. He is delighted by the catch: myriad flood-strewn sticks, a spring salmon, and four sockeye salmon. The spring salmon is large—at least nine kilograms of delicious meat—but the smaller sockeye are the real prize. "These are the first sockeye I've caught in years!"

I ask Jason whether he thinks the salmon got caught just before the landslide muddied the water, because the fish don't like to move in the dark. "Probably," he responds and resets the net.

The next day, the Bella Coola River remains dark and brown. There are no fish in Jason's net.

Jason is only 43, yet old enough to have seen his landscape transform. The mountains that surround the Bella Coola River and adjacent fjords are not the same ones he grew up with. When he was younger, the ridges up the inlet were covered in snow and ice year-round. Over the past three decades, these same mountains have become bare during summer. Entire glaciers have disappeared.

Jason and his father used to spend their summers together, fishing commercially for salmon. En route to the fishing grounds they would stop at Ice Box Bay and Little Ice Box Bay, on Dean Channel, where chunks of ice would drop from a glacier and roll down the mountainsides and into the water. They would pick up the floating ice and fill their fishhold with them. "We never had to buy ice to preserve our catch. The harbor did not even have an ice plant back then."

Today, when looking up from the Bella Coola Valley on a hot summer day, you can still see many glaciers. If you did not grow up here, you might look at the tons of ice still up there and think that things remain alright. But the mountains above this valley make me think of shifting baselines. Of longevity overfishing. Of old-growth forests replaced with young plantations. The shrinking of the massive, the elimination of the ancient.

Ice Box Bay and Little Ice Box Bay. Few young people have heard those names. The Bella Coola Harbor Ice Plant? Everyone knows that one these days.

Jason runs the Fisheries and Wildlife Lab of the Nuxalk Nation, conducting research that supports the management and conservation of salmon and bears. He and his sister Megan Moody—the former stewardship director for the Nuxalk and a biologist known for her research into eulachon declines—are among several Nuxalkmc who have been working with other scientists (myself included) to figure out why sockeye salmon that spawn in lakes draining into the Atnarko River, an upper tributary of the Bella Coola, declined precipitously in the early 2000s. A combination of changing ocean conditions due to climate change and large-scale fishery exploitation are the potential root causes of the stock decline, as they are for many other fish collapses. The strain is probably exacerbated by the increasing frequency of extreme and unpredictable high water, whether caused by sudden rains or by the rapid melting of glaciers brought by hotter springs and summers.

Melting glaciers are taking over my mind.

Part of it is nostalgia. In May of 1990, my friends Pierre Friele, Ken Legg, and I took a boat ride to the end of Dean Channel, about 60 kilometers north of the Bella Coola estuary as the raven flies. From there, we bushwhacked up to the snow and glaciated ridges, which we traveled for nearly two weeks on skis until we reached Kitlope Lake. We built a raft out of logs lashed together with our climbing rope and floated down the Kitlope River to its estuary at Gardner Canal. Seawater to seawater, over glaciers or on glacier-fed waters almost the whole way.

Sentimental memories aside, I want to understand how the loss of glaciers might impact Pacific salmon and potentially trigger another cultural loss for Coastal First Nations. If we maintain our current rates of fossil-fuel consumption, glaciologists predict that the glaciers of British Columbia's coastal mountains, including those in the Bella Coola area, will largely melt away by 2100, losing at least 70% of the volume and three quarters of the area they held at the start of this century. The forecast is even worse for the drier Rocky Mountains and other inland ranges of western Canada, where glaciers are likely to disappear almost entirely in the next 80 years. Given the momentum for warming in the climate system, even with an immediate decline in fossil fuel emissions we are locked into a long-term initial rise followed by a steady decline of meltwater flowing during late summer into glacier-fed watersheds like the Bella Coola River. Initially, until mid-century as the rates of glacial melt and water discharge escalate to a climax, it is likely that high water levels that muddy the water and destroy spawning habitat will be more common than they have been. Then, in the latter part of the century, as the glaciers dwindle and ice melt slows to trickles, low river levels and perhaps warmer water may challenge the ability of some salmon stocks to spawn successfully.

The loss of glaciers is likely to strain the link between people and bears—a relationship for which salmon serves as an intermediary. For the Nuxalk, their Heiltsuk, Kitasoo/Xai'xais, and Wuikinuxv neighbors, and for other Coastal First Nations, bears are relatives, part of family crests, honored through songs and dances. Many populations of black and grizzly bears, especially on the coast, depend on spawning salmon as their primary food during summer and early fall, when they are building fat reserves for winter hibernation. When there is not enough salmon, bears often compensate by turning to garbage, orchards, and other foods associated with people, which means salmon declines contribute to an increase in conflicts between humans and bears. This is not speculation. Ecologists recently estimated that losses of half the biomass of spawning salmon can lead to 20% increases in the number of bears shot to protect people in parts of British Columbia. The frequency with which grizzly and black bears amble through the town of Bella Coola seems to be increasing every year, which has prompted Jason and his crew of Nuxalkmc from the Fisheries and Wildlife Lab to build platforms on the river shore from which people can toss fish discards into fast flowing water, reducing incentives for bears to amble through the community and returning salmon nutrients into the river. They also have cut down brush that might hide bears at the edge of streets, thus preventing surprise encounters with people that might launch bears into defensive reactions. And some households have been putting electric fences around the sheds and smokehouses where they dry fish.

Yet some bears still wander into the community. People who are careless with their fish discards are an exception, but they do exist. And the town of Bella Coola has many fruit trees and shrubs—cherries, plums, raspberries—that attract hungry bears.

And all of this is happening while glaciers still exist. Where can we go from here?

Towards the end of my last visit to Bella Coola, Jason, his crew, and I wade up the Necleetsconnay River, which drains into the same estuary as the Bella Coola River. We look for spawning salmon but find none; early July is too early for this run. So I allow the aqua-blue water of the river's glacier-fed river to carry away my mind.

Jason and I come from vastly different cultural backgrounds. He is a Nuxalkmc rebuilding a culture after the catastrophe of colonialism. I am an immigrant from far away. Still, we have much in common. Both of us were close to our fathers, whom we admired for their values; both of them died of the same nasty disease, pancreatic cancer, the same year. Both of us have daughters with whom we are deeply connected. Both of us do what we do largely because we want to pass down to them the best world possible. Jason and I talk about these things every time we see each other, which works out to about twice a year.

Our common humanity. That is what keeps me going.

I know that it is too late to head back to where we came from. But we can surely head somewhere better than where we have been rushing to. Slow down, reassess, steer away from the cliff.

The Nuxalkmc lawyer Andrea Hilland writes, "The foundation of Nuxalk legal order is a holistic worldview that perceives the Nuxalkmc as having deep spiritual relationships with their territory and all living things." The Nuxalk origin story encapsulates these relationships:

> *Alquntam* [the Creator] created four supernatural Carpenters.... These beings...chiseled from wood a number of human beings, the forefathers of mankind.... The Carpenters did not confine their attention to men and women. Supernatural beings, animals, birds, trees, flowers, fish, mountains, rivers...all were created almost simultaneously....
>
> Around the walls of *Nusmata* [sky world] were hanging

> a number of bird and animal cloaks, representing ravens, eagles, whales, grizzly bears.... *Alquntam* asked each individual which of these cloaks he preferred to wear.... Each donned his choice from the wall and immediately became the bird or animal chosen. *Alquntam*...then sent him down in avian or mammal form.... [E]ach landed on the peak of a mountain in the Bella Coola country, took off his cloak, and reassumed human form. The discarded coverings floated back up to *Nusmata*.

Because humans removed their cloaks, Hilland explains, "the Nuxalkmc consider animals to be in a purer form than humans," which means that humans are devoid of superiority over other living things. Further, the first ancestors carried instructions from *Alquntam* on how to draw sustenance from plants and animals while simultaneously caring for them. These instructions came in the form of *smayustas*—"origin stories, names, songs, dances, and prerogatives"—which "convey lessons on appropriate use of Nuxalk territories and articulate consequences for breaching responsibilities."

The impunity of free markets, the broken democracies that support them, and religious world views that see humans as separate from biodiversity have certainly done their share of cornering us into our current global situation. But there is more to it than that. Individual choices we make daily, arriving at both constructive and destructive actions, count for a lot.

The economist Alfred E. Kahn introduced the notion of the "tyranny of small decisions" in the 1960s. As an example, he demonstrated how the choice of individual commuters to start using cars, planes, or buses indirectly terminated train service into Ithaca, New York. The interesting part is that most commuters wanted to keep the train, the only reliable way in and out of Ithaca during poor weather, but they unintentionally sabotaged it through their behavior. This is not unlike most of us today wanting a stable climate that would allow us to keep our glaciers while, contradictorily, making the carbon-intensive decision to fly to our holiday destinations.

When contemplating in the early 1980s how piecemeal management decisions exacerbate environmental problems, the ecologist William E. Odum wrote, "Unfortunately, it is much easier and politically more feasible for a planner or politician to make a decision on a single tract of land or a single issue rather than attempting policy or land-use plans on a large scale." Odum died in 1991, when awareness of human-caused climate change had barely begun to spread outside the realm of specialists. If he were writing today, Odum might have pointed out how the tyranny of small decisions has become inexcusable from a climate change perspective. To paraphrase Hilland, by now, "consequences for breaching responsibilities" have been amply articulated.

The greenhouse gases my personal activities emit during a single day would be inconsequential if they existed in isolation, but they do not. As humans, we make 7½ billion piecemeal contributions daily. It is true that per capita emission rates are orders of magnitude greater for the wealthy than for the poor, which highlights the social injustice of climate change. Still, if you add up the cumulative effect of individual emissions over multiple decades, our personal—small—decisions emerge as a default driver of our planet's condition. At the time of this writing, we have reached 415 ppm of CO_2 in our atmosphere, jolting the planet out of the stable, friendly climate from which we could have benefitted for many more centuries. But we still have a short window of time to turn these changes into something we can work with, if we choose to. The crucial question is, will we make this choice?

Science can illuminate the consequences of making that choice, or of failing to do so, but we need more than that. If we are not viscerally connected to what's at stake—from the perspective of a naturalist in love with the diversity of life, or of a hunter who derives spiritual and physical nourishment from the forest, or of a parent who exudes fear and love about intergenerational justice, or of cultures that have learned to reciprocate with Earth—then we are likely to see right past any of the science that can help us navigate towards a better Anthropocene.

One key element of the origin stories and legal systems of many Indigenous peoples is the belief that humans are only one component of biodiversity, rather than separate from and superior to it. As Robin Wall Kimmerer explains a fundamental principle of her Anishinaabe traditions, "We humans are the newest arrivals on earth, the youngsters, just learning to find our way." This perspective obligates humans to respect and learn from those beings who were here before us.

Yet these world views are notoriously absent in the collective psyche of industrial civilization, which, so far, has clung to the story that humans are separate from biodiversity. How fast we let go of that story, replacing it with others that portray humans as interrelated with other beings, will determine the fate of the biosphere for a long time to come.

This argument is not about the appropriation of indigeneity. Rather, it is about balancing political powers in ways in which Indigenous peoples become cocreators of the laws that govern our current civilization. This notion is beginning to take momentum in places like Canada, where Indigenous peoples are beginning to settle land rights, even land titles, and to work together with federal and provincial governments to protect lands and waters in ways that are consistent with their traditions and largely inclusive of settlers who came from faraway.

And as Kimmerer has argued, it is also about settlers behaving in ways that respect biodiversity and follow the example of *Nanabozho*, the Original Anishinaabe Man, who upon arriving on Earth understood that "this was not the 'New World', but one that was ancient before he came."

7

Beautiful Protest

It is late April of 2014. The morning is calm, cold, and drizzly. I am leaving Klemtu with Ernie Mason, Charlie Mason's son and a key player in the fisheries program of the Kitasoo/Xai'xais Nation. Thankfully, *Cidu*, the eight-meter-long powerboat that belongs to the fisheries program, has a cabin where we can shelter from the chill and dampness. We enter Mathieson Channel, where eight harbor porpoises play in our wake for about a kilometer as we begin to travel north. At the narrows known as Hell's Gate, seven Steller sea lions, including two large bulls, have hauled themselves out onto a ledge a few meters above the water, where a steep forest descends to the edge of intertidal granite. We slow down momentarily to have a look; their sonorous barks and strong, fishy smell invade my senses. Shortly after, we enter Mussel Inlet. Steep granite walls rise out of the water, and the surrounding mountains still hold most of their winter snowpack. At sea level by the estuary rests a large patch of deep snow—the deposition zone of avalanches released from the gulley above by the vagaries of weather and the precision of gravity. I look intently at the estuarine meadow, hoping to see grizzly bears digging for roots to eat. None today.

Yet shortly after, a mere thirty meters above the inlet, we see four mountain goats on the mossy ledges of a cliff adjacent to a waterfall. Ernie has a long family history of connection to these animals.

His paternal great-grandparents were among the last inhabitants of Q'enxv, a village that once nestled in the adjacent fjord, Kynoch Inlet, where mountain goat hunting was a vital part of the people's seasonal round. And Ernie learned to continue that tradition while hunting with his father.

On the surface of the bay ten floats bob, spaced 100 meters or so apart. Ernie and I had set them the day before. A line runs from each float to a crab trap that rests on the bottom. Ernie maneuvers *Cidu* to the first float, grabs the line and runs it through an electric hauler that pulls up the trap. Inside are several Dungeness crabs, which we count, measure, identify their gender by the shape of their abdomen, and score for other biological features before returning them, alive, to the water. We repeat the protocol for each of the remaining traps. By afternoon we are at the estuary of the Kynoch River, where we do it all over again. Over the course of several days, we will sample Dungeness crab at four additional sites.

Roughly at the same time, give or take a week, the three other First Nations of the Central Coast are performing the same tasks in their areas. Collectively, the Kitasoo/Xai'xais, Wuikinuxv, Heiltsuk, and Nuxalk will have covered 20 research sites spread over 11,000 square kilometers. They will repeat the sampling every two months or so, pausing for the height of winter, for years to come.

As we head back to Klemtu, I am giddy with the prospects of big science, of big law.

Scientists like to measure everything relevant to their studies, attaching precise numbers to objects, organisms, energy, time, distance, and relationships. That is a staple feature of our craft. No quantification, no science. Ecologists strive for synthetic understanding of multiple pieces of information; to do so, we rely heavily—though not entirely—on powerful computers, mathematical theory, and statistics. Yet a focus on precise numbers can provide insights that, depending on context, are limitless, limiting, or somewhere in between.

People who rely on traditional knowledge also quantify their observations. But they do so in simpler ways that often are more practical and, though less precise, are just as or more accurate than those used by ecologists. (The distinction between precision and accuracy is important. More precise measurements are less variable and more detailed, while greater accuracy better reflects the true nature of what is being studied. The two need not be mutually exclusive but can be.) For such people, the process may take on a more diffuse quality that yields insights that are no less "true" than those produced by Western science. Traditional fishers and hunters assess the relative abundance of their prey—in specific habitats at specific times of year—based on how frequently prey are caught in snares, nets, and traps, or encountered as people travel known distances or sit still for known periods of time. For hunters, prey encounters need not be direct: they may take the form of tracks, marks on vegetation, and smells left by animals. Similarly, fishers may indirectly discern the underwater presence of fish schools, which, in the words of the writer Anna Badkhen, can manifest as "an indentation in the surface, an irregularity in the wave pattern, a boil of bubbles you can see even in the blowing water of a gale." People may or may not measure time and space with watches, charts, and GPS devices. The quality of a familiar journey—between regularly used camps or fishing sites, in good or bad weather—approximates spacetime.

In some cases, the right time to harvest a target species can be read from parallel signals from other species. Ethnobotanist Nancy Turner describes how, in Coastal British Columbia, "the full expansion of alder leaves, the flowering of salmonberry," or "the growth of stinging nettles" observed at the village "predicts the stage of readiness" of edible seaweed growing annually on the intertidal zone of rocky shores. From these signals people gauge when growth has nearly peaked: the "optimal time" to leave the village, head out on boats, and harvest seaweed before it begins to senesce.

To assess body condition of their prey, hunters often rely on qualitative signals. For example, Cree and Dëne hunters from northern

Canada or Alaska estimate the fat content of live caribou based on the shape of the animal's rump and the quality of their gait. When combined with observations on the abundance of young caribou in proportion to that of adults, these signals allow hunters to eyeball whether prey populations are likely to increase or decrease. Traditional plant gatherers make similar assessments of plant populations from the quality and relative abundance of berries and flowers, and the extent of new growth.

As anthropologist Hugh Brody documents in his book, *Maps and Dreams*, Dunne-za hunters of Northeastern British Columbia *dream* to ascertain the state of animal populations and whether it is appropriate to carry out a hunt. For those who adhere narrowly to Western science, this notion might seem implausible. But the fact is, whatever mechanism underlies the dreaming—a subconscious processing of detailed information already stored yet inaccessible to the conscious brain, or otherwise—it works. Traditional Dunne-za hunters find their prey when they choose to, and they have self-regulated their hunting in ways that have allowed them to live sustainably in their territories for many centuries. Not all traditional resource gatherers rely on dreams to make critical decisions, yet most use some intuitive approach imbued within their cultural context to interpret complex and multiple lines of evidence about the state of ecosystems.

When synthesized in the mind, the multiple pieces of information that traditional hunters, fishers, and gatherers have obtained—through observation, inherited knowledge, dreaming—determine their decision to kill an animal, gather a plant, or not. The ability to observe and interpret measures and signals on the move and without special instrumentation allows networks of socially and culturally connected fishers, hunters, and plant gatherers to continuously collect information over vast areas, regularly integrating and updating it into a combined assessment of their shared ecosystems. In contrast, many of the more quantitative measures of Western science—important as they are for their precision and potential for detailed predictions of ecosystem change—can be difficult or expensive to obtain in the field, which means that they often are fragmented in time and space, or not collected at all.

Scientists are trained to be skeptical and follow a narrow set of principles to reject or tentatively accept a hypothesis, and many need evidence of common ground before they can open their minds to the value of traditional knowledge. To that end, Western ecologists Henrik Moller and Fikret Berkes have played significant roles in testing the extent to which observations by traditional resource gatherers can be applied to modern resource management. Based on their large body of work, these scientists argue that, almost always, there is great agreement between traditional ecological knowledge and science. That message has begun to spread among resource managers and others trained in Western science, but it takes time to absorb.

As for holders of traditional knowledge, to them, the longevity and sustainability of their place-based cultures is ultimate proof that their understanding of ecosystems is real. They do not need Western science to validate their insights.

There was a time, not long ago, when people from Bella Bella, Klemtu, Bella Coola, and Wuikinuxv Village would board a rowboat or a small skiff, travel to a nearby bay or channel, and drop into the shallows a tethered, homemade metal hoop with a strung net and a fish carcass attached as bait. They would then watch Dungeness crabs walk onto the hoop trap. Through clear water, they would size each crab. If it was large and, almost certainly, an adult male that had already reproduced for several years, traditional law allowed them to harvest it; if it was small and, very likely, a female or a younger male, they left that crab in the water. They would also watch other crabs walking along the bottom and, based on prior experience of the place, assess whether local crab numbers were going up or down, or were stable. Then they would think about relatives back home. Who has not had crab for a while and could use some? How many do I need for my own household? Do we have enough for the next potlatch? All these pieces of information would then determine whether the harvesters would haul the hoop trap with crab or empty, and whether they would reset it.

During that time—which elders who fished from the 1930s and into the 1990s say lasted most of their lives—everyone could find the crabs they needed. And that was a good thing, not only for food but also for culture. A potlatch without Dungeness crab is a table missing one of the guests of honor. But nobody worried about that back then.

Change began in the late 1990s. Cage traps, which crabs enter through one-way gates, and larger boats with better engines, became more available in the villages, allowing people to harvest crab farther afield and at deeper depths. That, in itself, did not harm the Dungeness crab population. People maintained their fundamental behavior, keeping a mental tally of crabs caught per trap during each fishing trip, regulating their harvest according to traditional laws. For almost everyone, these practices still endure.

But, at the same time, exploitation by commercial and recreational fishers increased and kept ramping up into the 2000s. Indigenous fishers observed a drop in the number of adult males caught per trap. Eventually, they were catching less than one quarter of the usual number of adult male crabs.

Law is law. So the people stopped fishing places that were hit hard by commercial and recreational fishers. With fewer harvesting areas available, the presence of Dungeness crab at potlatches stopped being taken for granted.

The four First Nations of the Central Coast are one another's relatives in many ways. As you might expect for any family, there are challenges in their relationships, but what stands out is the way people from different communities look after one another when times are tough. Over the past half decade, I have witnessed these Nations working together towards their common goal of restoring crab populations, which has taken them through some major challenges. In the process, I also saw a shift in the zeitgeist of DFO crab managers, who dramatically improved their willingness to work with First Nations in the collaborative management of culturally significant species. I believe this shift symbolizes something bigger, something that transcends the Central Coast of British Columbia—a sign that, globally, it may be possible for governance and resource management systems to become inclusive of Indigenous peoples and their wealth of knowledge about ecosystems.

In British Columbia, DFO manages Dungeness crab populations by allowing only the harvest of large males with carapace widths of at least 165 millimeters. By the time they reach that size, most males have been reproducing annually for at least a couple of years, so managers assume that if females and smaller males are not harvested, then reproductive rates remain adequate and the fishery is sustainable. Managers are so confident of this assumption that they rarely collect monitoring data to corroborate it. In a theoretical sense, this is not necessarily flawed, especially if governments underfund their fishery scientists—as Stephen Harper, Canada's prime minister from 2006 to 2015, notoriously did—forcing managers to triage scarce resources for monitoring. But when First Nations' observations began to contradict the regulatory agency's theoretical (and unverified) assumption, DFO crab managers proceeded to ignore Indigenous voices for the next 10 years.

By 2007, the Wuikinuxv, Kitasoo/Xai'xais, Nuxalk, and Heiltsuk were frequently approaching DFO as a unified voice, requesting the closure of commercial and recreational fisheries at their traditional fishing sites. Although Canada's constitution protects traditional Indigenous fisheries, giving them precedence over commercial and recreational catches when a resource is not abundant enough to support all three, DFO initially denied these requests, arguing that there was no evidence for a conservation problem or need for management intervention. In their view, observations by place-based fishers were not valid enough to contest "scientific" criteria used by federal managers.

By the time I started working with Central Coast Nations in 2013, interactions with DFO had become combative. "Prove that you have a problem," had become the tired, and vacuous, challenge from one particular crab manager.

So the Nations turned to their traditional laws.

In early 2014, leaders from Central Coast Nations closed a total of ten bays or channels to commercial and recreational fishers. The closures were implemented under the legal principles that require hereditary chiefs to protect resources within their tenure areas. The Nations asked DFO crab managers to recognize and legislate the closures into fishery regulations. DFO denied the request.

The Nations then invited the cooperation of commercial and recreational fishers, who generally complied, and instructed me to use the closures to conduct a large-scale ecological experiment to test for fishery impacts on crab populations. To carry out the experiment, we chose ten additional sites where all fishers could continue their activities, giving us points of comparison for the sites that were closed. By the spring of 2014, teams from all Nations where on the water, collectively monitoring the 20 sites and ensuring that commercial and recreational fishers stayed clear of the closures.

Over the next 10 months, adult male crabs became more abundant and larger at closed sites, yet smaller and fewer at sites open to commercial and recreational fishers. Females and subadult males, which are illegal to harvest, maintained similar abundances at closed and open sites. These results implicated commercial and recreational fishers in the declining catch rates Indigenous fishers had experienced.

When we returned to the negotiating table with DFO, I expected our experimental data to settle the issue. How naive. DFO managers did not dispute that our results showed a fishery impact. But now they argued that we had yet to demonstrate that Indigenous fishers were having a problem catching enough crab to satisfy their cultural and food needs. In addition to an obvious disregard for the precautionary principle, by which the federal agency claims to abide and which guides traditional Indigenous law, the implicit message from DFO crab managers remained racist: observations by First Nations were irrelevant, and concern for cultural fisheries was secondary to that of commercial and recreational fisheries, no matter that the Canadian constitution stated the opposite.

By 2016, the Nations threatened to sue DFO. I was summoned to a law office to sign an affidavit about our published research, which was to be used as evidence in court.

At the same time, two colleagues from the University of Victoria, Lauren Eckert and Natalie Ban, joined our effort to translate the plight of Indigenous communities into something quantitative that might break the ice and establish the common ground needed to work constructively with DFO. Through interviews, Lauren and Natalie reconstructed the fishers' past histories of harvesting and lifetime observations of crab populations in relation to commercial and recreational exploitation. Some of the people they interviewed related six decades of observations.

The story that emerged was simple and powerful. During the decades that preceded the late 1990s, fishers consistently caught, on average, 22 crabs per trap. At the time of the interviews, a typical catch averaged five crabs per trap. Perhaps more importantly, people defined a threshold for the minimum catch rate that—based on their expectations from lifetime experience—characterizes a successful harvesting trip: 15 crabs total, caught with two traps. Our monitoring data quantitatively demonstrated that meeting that threshold was improbable at almost all sites we had studied. This was the evidence—the "proof"—that people were having a problem accessing an important traditional food. Meanwhile, the elected chiefs had been meeting with high-level directors from DFO, demanding that the federal agency meet its legal obligation to protect traditional fisheries.

In 2017, reason began to prevail. Discussions and negotiations between Central Coast First Nations and DFO became constructive, producing a joint technical working group—involving First Nations, their scientific staff, and DFO—dedicated to the recovery and management of Dungeness crab in the Central Coast. To date, the outcomes include four closures for commercial fisheries, which honored requests by First Nations to protect Dungeness crab in their territories and were federally legislated in 2017 and 2018.

This shift provides tremendous grounds for optimism. It heralds the potential for rapid change in the right direction; for the chance of coalescing different ways of knowing—and different legal systems—into a more sustainable relationship between people and the Earth. It is, too, a clear sign that First Nations' leaders have no patience for a "victim narrative."

"I do not feel colonized," Doug Neasloss, Chief Councillor and Stewardship Director for the Kitasoo/Xai'xais Nation, said to me one evening in September 2018 after I had made some statement about the lingering impacts of colonization. So I put my tail between my legs, acknowledged that mild scolding, and kept listening.

Currently in his mid-30s, Doug has been the political leader of his people, almost without interruption, since he was 27, and many of his efforts have focused on cultural rebuilding and the empowerment of youth. Some efforts have been proactive, but many have been a response to the alcohol abuse, violence, and self-harm that have been part of the aftermath of colonization in many First Nation communities. But Doug, like many other Indigenous leaders, does not linger on those ugly elements. Instead, he focuses on the fact that First Nations are regaining their political strength. That they have established a precedent for closing commercial fisheries in their territories. And he envisions a near future in which First Nations lead the establishment of marine protected areas that safeguard their traditional harvests and biodiversity and that are recognized and supported by the federal government.

Later in the night, we laughed about how the crab closures shook things up. "That crab study was a beautiful protest," Doug said.

Protest.

Cooperation.

Cultural empowerment.

Social justice.

Climate justice.

A better Anthropocene.

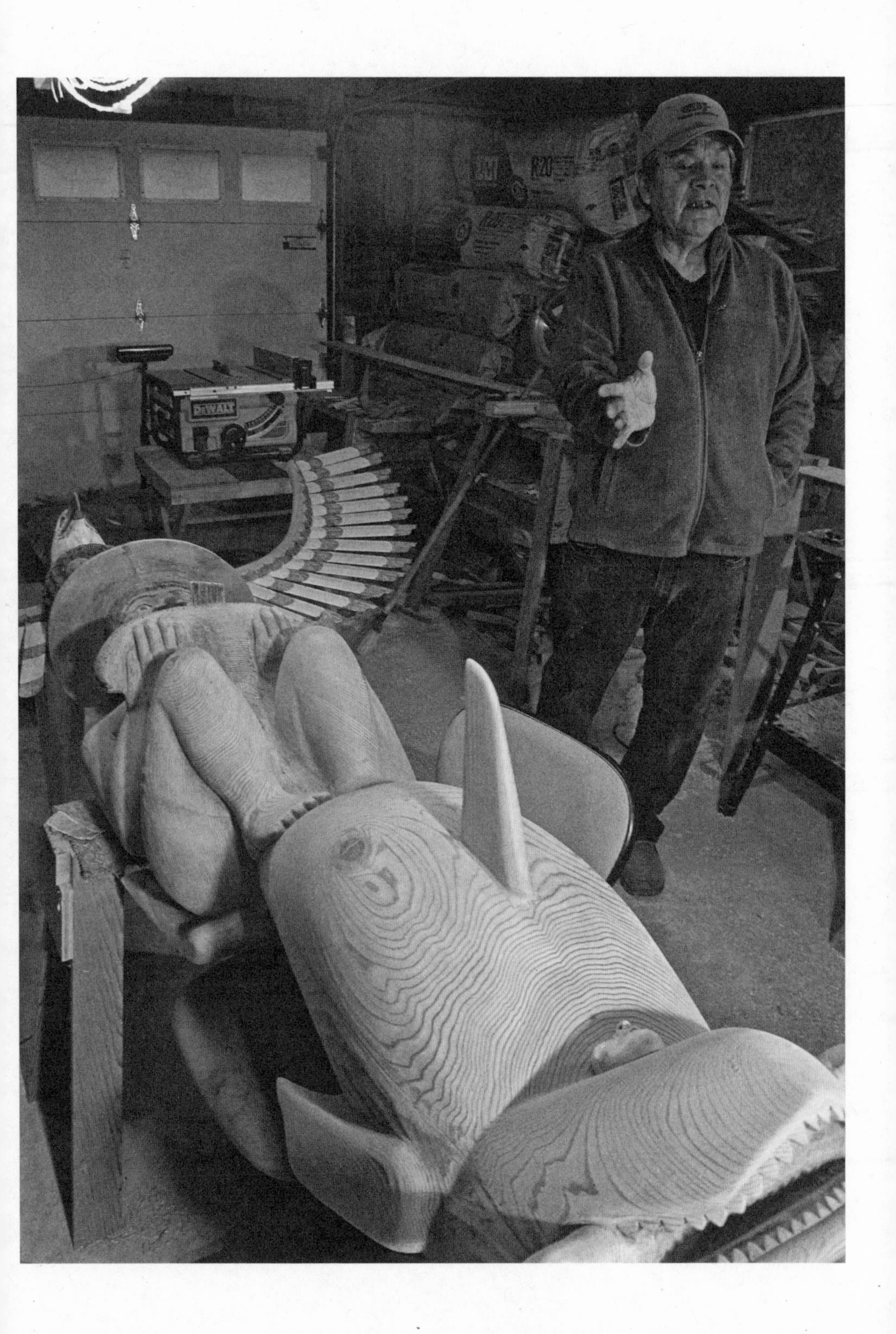
R20
DEWALT

Interlude I

The cabin heater of *Cidu* is not working. Harsh timing.

I wake up in Klemtu to a clear morning of March 2018. The receding tide has left a blanket of ice on the beach; the freshwater layer at the surface of the bay has frozen too. Snow covers the forested mountains almost down to sea level.

Yet the winds remain calm, and Ernie Mason, Tristan Blaine, and I head out on *Cidu* to look for spawning herring and to conduct dive surveys for rockfish off the west coast of Aristazabal Island. We leave the fjords and enter the outer coast. Swells rise and drop against the cliffy shores of tiny islands. By midmorning, rain clouds move in from the southeast, raising the temperature above freezing, but just barely.

We spend the day in and out of the ocean. The rain turns torrential.

On the long trip back to Klemtu, I wear every stitch of clothing I can get my hands on. In a feeble attempt to keep up my body temperature, I move around as much as I can within the confined space of the unheated cabin.

While crossing Kitasu Bay, we see Charlie Mason with a crew of three 30 to 50-year-olds (what Charlie calls "young guys") hauling up a halibut longline from his open punt. Ernie maneuvers *Cidu* alongside his father and, through an open window from inside the cabin, jokes with him about how we better rush home because our heater broke.

Seventy-four-year-old Charlie Mason, hereditary chief Neasmuutk Haimas, dressed in the full regalia of rubberized raingear, lets go of the longline while instructing the young guys to keep hauling it. Grandly, he raises his arms and gaze towards the frigid rain that is pounding him and everything else outside *Cidu*'s cabin. Then he stares into every section of his roofless vessel, turns back to us and pronounces:

"My heater broke too."

8

Echoes Across the Lake

I spend the night by Konni Lake. Out in the open, cocooned in my sleeping bag, sheltered from the autumn cold. The vastness of stars shines in extreme clarity. The waning moon dilutes some of the starriness, but not too much. Just enough to create a gradient of luminosity into the seeming infinite.

Konni Lake is in the Nemiah Valley, on the east side of the Coast Mountains of British Columbia. If I were to go over the ridges, cross some glaciers, and drop down the west side, I would connect with the Central Coast, where I spend so much of my time working. Yet tonight I lie down within the territory of the Tŝilhqot'in people—archetypal hunters and defenders of tradition. Their stories speak of ancestry, connection to place, the marrow of ungulate bones. From battles with colonial surveyors 150 years ago to modern court cases against mining and logging companies, they are fierce protectors of salmon, trout, forest, bear, moose: the building blocks of their culture.

The previous year, on June 26 of 2014, the Supreme Court of Canada granted the Tŝilhqot'in people victory against British Columbia. Tŝilhqot'in Nation v. British Columbia redefined Aboriginal title, which up until then had been constrained to small areas of year-round residence: reserves where people had been forced to aggregate in the aftermath of colonization. For the Tŝilhqot'in people, Aboriginal title now applies to 1,700 square kilometers of lands historically and currently used for hunting, fishing, trapping, plant-gathering, and other cultural practices. That means that logging, mining, and other resource-extraction companies can no longer conduct business in the titled area unless the Tŝilhqot'in people grant explicit permission. Though the court ruling does not apply to other First Nations, or even to all the ancestral lands of the Tŝilhqot'in people, this was a landmark case. The morning of the ruling I was in the coastal village of Bella Bella, about to meet with hereditary chiefs and technical staff from the Heiltsuk Nation, when Kelly Brown, the stewardship director, walked into the room with a big grin in his face and declared, "Today the Indians got a little stronger."

A glimmer of hope that the world can become less racist.

I wake up to a frosty, foggy morning. Laze around in my cocoon until the sun breaks out and ignites the yellow leaves of quaking aspens on the hillsides. A little stiff and cold, I gather beach wood and light a fire. Warmed by its roaring flame, I sip yerba mate, writing these words. Moon, still high in the sky. The sun rises above the mountainous horizon and dissolves the mist, everywhere except for a thin layer over Konni Lake. Five common loons emerge through that vaporous blanket. I have been hearing their songs throughout my writing. Coyote howls, chickadee whistles and the calls of bald eagles fill the spaces in between.

I take it all in, feeling grateful for my life.

I have always been conscious of having an intentional path. I have not always succeeded. Failed miserably at times. Yet it is hard to imagine a more beautiful life than the one I've been living. Grateful beyond words for my wife Gail and my daughter Twyla Bella. Thankful for the opportunity to apply my training in ecology, my connection to the land and sea, to help Coastal First Nations, and the rest of us, navigate through a rapidly changing planet in the face of biodiversity loss and climate disruption.

That is what this journey into Tŝilhqot'in territory is all about. Honoring my obligation to ground by closing the door, temporarily, to the barrage of professional responsibilities—no matter how important and stimulating—that explode on my face whenever I enter my work space.

I am on my way to join a hunting camp of the Tŝilhqot'in First Nations at Teztan Biny, a subalpine lake the Tŝilhqot'in consider sacred. I am traveling at the invitation from Marylin Baptiste, former Chief Councillor of the Xeni Gwet'in, one of six Tŝilhqot'in bands. I first met Marylin in 2010, when Twyla Bella, who was six years old at the time, and I joined her and others on a canoe journey that drew attention to the plight of wild salmon in the Fraser River watershed, including its upper reaches in Tŝilhqot'in territory, which have been facing mounting pressures from extractive industries and climate change. That year, in her position as chief councillor, Marilyn had written in the op-ed pages of the *Globe and Mail*, a national newspaper:

> What would you do if another country, many times more populous and powerful, decides that it wants Canada's water and, after listening to all the reasons why it cannot not simply take it, announces that it is going to do exactly that?
>
> Would you refuse to accept the country's justification that its hundreds of millions of people desperately need the water to sustain their economy and that this outweighs any harm that would be done to the relatively small Canadian population that stands in the way? Would you expect your governments to resist? If your answers are yes, then you have an idea of the position of the Tŝilhqot'in people.

Marylin was referring to the proposal for a copper-gold mine that, if developed, would destroy Teztan Biny and a portion of the traditional lands of Tŝilhqot'in Nations in the name of a projected C$5 billion worth of economic activity over 20 years. She explained that "gouging out a 35-square-kilometre open-pit mine would kill a lake, streams, wildlife, forests and our rights and way of life," concluding that "no amount of money could compensate for what this mine would do to our people and the land we treasure."

Teztan Biny is just outside the area where Canada's highest court recently recognized Tŝilhqot'in land title, so it remains under threat by corporate forces. Five years after Marylin's op-ed, the mining permit has not been granted, but it also has not been denied outright, heightening the fact that places like Teztan Biny will remain vulnerable until the industrial civilization begins to collectively embrace an alternative story.

En route to Teztan Biny, I make a detour and drive the short gravel road between Konni Lake and the Nemiah reserve. I park and walk through "downtown" Nemiah. I am on a mission to track down Edmond Lulua; we first met two years ago and did some walking around Nemiah. Edmond is always keen to head into the bush—to hunt or for the sheer pleasure of being on the land. This time I need his help as a translator. Edmond is about 50. Like most of his generation, he was introduced to English when he was sent to residential school, at the age of seven or so; for people his age, Tŝilhqot'in is the first language. And many of the older people who spent their entire childhoods in the bush, untrackable by police and priests rounding up children, avoided the schools and never learned English.

I find him at home, at the end of a short dirt track. Edmond is tinkering with tools and firewood inside a shed beside his modest house. I explain what I am after and, within moments, he points to two bikes leaning against the shed. We hop on the bikes and start pedaling up a gravel road towards the log cabin where Julianna Lulua lives with her husband, Ubil.

Julianna is an elder whom I first met in May 2014, harvesting traditional food plants with a group of local and visiting school kids, among them my daughter. On that beautiful spring day, kids, adults, and elders walked on subalpine meadows overlooking the turquoise waters of Chilko Lake and its surrounding snow-covered mountains, summoned by the small white flowers of the perennial plant known as *sunt'iny* to the Tŝilhqot'in and *Claytonia lanceolata* to Western botanists. With sticks, metal bars, and deer antlers we dug for *sunt'iny's* nutritious corms—the "mountain potatoes" that we would later steam into delicious meals. Julianna had led her younger relatives in a steady harvest, bending over one digging spot after another, until her plastic pail filled with fleshy, energy-rich gifts from the earth.

Later that summer I met Julianna again, this time by the Chilko River, an upper tributary of the Fraser River watershed, where the six Tŝilhqot'in bands had gathered to gaff salmon, share songs, and play traditional games. At that gathering I watched her, along with her son Stanley and daughter Nancy, butcher two huge mule deer that Stanley had recently shot. I have hunted many deer and pride myself in being good at field dressing and butchering game, but what I witnessed that day was a completely different skill level. Within half an hour, Julianna, Stanley, Nancy, and a couple of other helping hands skinned, processed, and hung in the smokehouse the meat of both bucks. Their movements were fluid, efficient. No one got into another's way. At one point, Stanley used a hatchet, delivering precise, gentle blows to quickly sever bone joints and tendons, separating each hind leg from its body. With their permission, I photographed those moments and their aftermath: Julianna standing proud in the smoke shack, surrounded by hanging fillets of salmon and deer hind legs—the bounty of the rivers and land from which she and her people are inseparable.

Those photos made it into my first book, which—as I chase on a bike after Edmond—is fresh off the press. I want to thank Julianna by presenting her with a copy. She does not speak English, so I need Edmond to translate.

We reach Julianna's log cabin, drop our bikes by the entrance, knock and walk in. I am nervous about visiting without an invitation, but Julianna's welcome quickly puts me at ease. Edmond and I sit beside Julianna, and Edmond guides her through the book. She loves her portrait in the smoke shack. Edmond then browses through the book to show her photographs of my marine research and gets me to explain my work with Coastal First Nations.

"*Nexwachanalyagh Gulin*," Juliana says when she bids us farewell at the door: "We thank you very much."

I arrive at Teztan Biny in the afternoon. A small group of people from several Tŝilhqot'in communities have been camping here for several days. The camp has a spiritual focus, a place for people to reconnect to the land and to heal from substance abuse through hunting, sweat lodges, and ceremony. The camp is the brainchild of Cecil Grinder, of the Tl'etinqox band. He and I have not met before. Yet over the next few days we develop a friendship that will keep drawing me back to Teztan Biny, year after year.

Cecil is in his late 50s. Tall and fit, he carries himself with tremendous confidence while exuding the kind of energy that invites you to be as honest with him as you can. These qualities must have served him well during his 20 years as officer of the Royal Canadian Mounted Police, mostly in the town of Williams Lake, the largest settlement in the area and a hub for outlying First Nation communities.

Cecil was brought up largely in the bush, speaking Tŝilhqot'in as his first language and hunting and fishing with his maternal grandparents. When he was an infant, his mother left his father, who had turned into an aggressive alcoholic. When Cecil started drinking and hanging out with minor criminals as a teenager, his mother marched him into the local police detachment, where he agreed to be recruited. Cecil credits his mother's determined move with the transformation that, over the years, changed him from one who needed help healing into one who helps others heal. Now retired from the police force, he dedicates himself to that task, and to the strengthening of Tŝilhqot'in culture.

I ask Cecil if he ever met his father again.

"Yes. On the job. Arresting him."

Squalls of wet snow. Truck stuck in mud, rescued when we jack it up and jam sawed logs under the tires to create traction. Disaster forestalled.

About a hundred sandhill cranes vee past us.

Edmond Lulua's brother Alex and I are looking for moose. He spots one through the sleet: a bull melting in and out of distant brush. I shoot. Miss. Alex shrugs in a way that appears to convey, "I live here and will get one tomorrow or the next day," which stops me from feeling like crap, but just barely.

By a remote lake in the area known as Nabaŝ, near the remains of log cabins where Xeni Gwet'in families lived, year-round, well into the 1970s, Alex makes a moose call. A pack of wolves howl back.

We return to camp as others enter the sweat lodge. We join them.

Rounds of hot steam, drumming and song for our female ancestors, for our male ancestors, for ourselves, for our other relatives and the rest of the world. Between each round: a walk through the frosty air and a plunge into Teztan Biny.

Under the Milky Way.

Wake up to a crystal-clear frosty morning. Walk around Teztan Biny with Jarvis, 19 years old. No moose, but we shoot three grouse. He tells me how he was sober only three days of the previous summer. How he ended up in court after a high-speed cop chase when he was driving drunk. How he is able to hunt moose here today only because he got out on parole, on condition that he test weekly for substances in his blood. Good hunter and so much potential. Back to camp. Feast on the grouse we brought back and the trout the others caught. In the evening, another sweat lodge. The stars. The cold lake.

Another clear and cold day. I am hunting moose with Kenny William, who is roughly my age. We walk and scan the landscape through binoculars until I spot a large bull standing at the northeast end of Teztan Biny. Hopes soar. But the bull is over a kilometer away; for the next hour or so we fail to call it in within shooting range or to track it down.

During a break, Kenny tells me how he has had headaches, continuously, for more than 40 years. Ever since a teacher at school punched him with a fist on the head. Standard educational technique during the Canadian apartheid. The assault probably gave Kenny a concussion that was never diagnosed. Stunned him for decades. He continues to take refuge in substance abuse.

Yet here he is. Hunting. Trying to heal.

Back at camp, Marvin William, Kenny's father and a most respected hunter, is about to go fishing on a small rowboat. Kenny joins him. The Tŝilhqot'in language flows between father and son like song.

The previous night, a young man named David, who has not learned the language, had remarked, "I like hearing the elders talk in Tŝilhqot'in. It soothes my spirit."

By afternoon, dozens of people from the Tŝilhqot'in communities join us to commemorate the one-year anniversary of the raising of a very special pole. Carved in red cedar and gifted to the Tŝilhqot'in peoples by the great Nuu-Chah-Nulth artist Tim Paul, the pole symbolizes the struggle to protect Teztan Biny from the corporation that wants to mine it. And more. The Nuu-Chah-Nulth people are from what is now known as the West Coast of Vancouver Island, some 300 kilometers to the southwest of Teztan Biny. Marine waters, rugged mountains, and many glaciers lie in between the two places. Paul's masterpiece, therefore, encapsulates the spirit of cooperation that pervades relationships among many modern First Nations—including those Nations that, in pre-colonial times, may have had little or no direct contact with each other.

Indigenous carvers often are great orators, and Tim Paul is no exception. The previous year at this very spot, during the pole-raising ceremony of 2014, eloquent silences had amplified the strength of his words:

> The spiritual power that is present here, we feel. I feel. And we are here to give you what we have.... Our grandparents are gonna take root here. To build with you strong support and alliance.... We come up here and will distribute our wealth.... We distribute our wealth in a faraway place.

And now, one year later at the same lakeshore, Cecil renews the ceremony.

Cecil beats his drum, punctuating the pauses between his strong and eloquent words on the sacredness of Teztan Biny. The surrounding mountains are covered with a dusting of snow. The drum reverberates across the water—three beats in a row followed by three thunderous echoes. The first two echoes are relatively quiet, yet the final one is much louder; it seems to emanate from the direction of Ts'ilʔos—the dominant mountain to the southwest of Teztan Biny.

Ts'ilʔos once was a man. He still watches over the territory.

Cecil's elocution is directed as much at the people as it is at the isolated dark clouds surrounded by blue sky. With each successive drumbeat, darkness seems to shrink. Blue sky appears to expand.

Indigenous peoples do not need the rest of the world to initiate their healing. What they *do* need is a just society that cares for the land, fish, animals, and plants that make everyone whole. The question is whether the rest of us will choose to live up to our potential of wisely practicing reciprocity, with each other and with our non-human kin.

To shift away from the path that threatens all ecosystems and our livelihoods, the most important thing we can do is connect with the Earth and one another in meaningful ways. Everything else that matters will follow. I believe this to be especially and increasingly true for the scientists engaged in the hard, objective work of saving us from ourselves. If scientists—who are only people—do not connect with the Earth and other humans in meaningful ways, then we may fail to ask important questions about the consequences of alternative policies for reducing emissions (or not) and mitigating biodiversity loss (or not).

Who benefits and who suffers and how badly? Which combination of plants and animals—and the myriad relationships among them—get to endure or go extinct? Which human cultures—and their unique world views that allow alternative ways of seeking meaning to coexist—get to live, or die?

True, many scientists already understand that the questions we ask are inextricably linked to our connections with place and people. You can see such awareness in some brilliant studies already out there, particularly those that analyze feedbacks between cultures, economies, and ecosystems. Yet, for the most part, that awareness is too vague. It needs to strengthen, become more widespread and explicit, so that scientists may provide better guidance on how to navigate the rapid changes that our world is undergoing.

The scientific method is objective in the ways it seeks answers to the questions we ask. We make an initial observation, form a tentative explanation that we call a hypothesis, and then test that hypothesis against the more formal and systematic observations that we call data. If the data do not contradict it, we continue to tentatively accept that hypothesis, while continuing to scrutinize our ideas against other evidence and alternative explanations. That is about as objective as humans get.

But while the methods that scientists use to answer questions are objective, the individual world views that ask these questions are not. If we have never witnessed a potlatch ceremony, where young and old connect to ancestry through the traditional foods that are served and acknowledged—herring eggs, halibut, seal, seaweed—then we may never ask how the combined effects of industrial fisheries and climate change indirectly damage the social fabric and cultural integrity of Indigenous peoples. If we have never walked the forest with a 19-year-old on parole whose tracking skills get us closer to the moose we seek to hunt, then we may never ask how a mining project that disrupts the lakes and forest may take away the places of cultural refuge that heal traditional hunters in the aftermath of colonization. And if scientists fail to ask these sorts of questions, then policy makers and politicians will lack the data to understand the true consequences of their decisions.

So yes, I am here to ground. That is my obligation.

Pelican 15.5

9

Ditching Our Climate-Wrecking Stories

The climate and biodiversity crises reflect the stories that we have allowed to infiltrate the collective psyche of industrial civilization. It is high time to let go of these stories. Unclutter ourselves. Regain clarity. Make room for other stories that can help us reshape our ways of being in the world.

For starters, I'd love to let go of what has been our most venerated and ingrained story since the mid-1700s: that burning more fossil fuels is synonymous with prosperity. Letting go of that story shouldn't be too hard these days. Financial investment over the past decade has been shifting very quickly away from fossil fuels and towards renewable energies. Even Bob Dudley, group chief executive of BP—one of the largest fossil fuel corporations in the world—acknowledged the trend, writing in the *BP Statistical Review of World Energy 2017*: "The relentless drive to improve energy efficiency is causing global energy consumption overall to decelerate. And, of course, the energy mix is shifting towards cleaner, lower carbon fuels, driven by environmental needs and technological advances." Dudley went on:

> Coal consumption fell sharply for the second consecutive year, with its share within primary energy falling to its lowest level since 2004. Indeed, coal production and consumption in the UK completed an entire cycle, falling back to levels

> last seen almost 200 years ago around the time of the Industrial Revolution, with the UK power sector recording its first ever coal-free day in April of this year. In contrast, renewable energy globally led by wind and solar power grew strongly, helped by continuing technological advances.

According to Dudley's team, global production of oil and natural gas also slowed down in 2016. Meanwhile, that same year, the combined power provided by wind and solar energy increased by 14.6%: the biggest jump on record. All in all, since 2005, the installed capacity for renewable energy has grown exponentially, doubling every five and half years.

It is important to celebrate that King Coal—that grand initiator of the Industrial Revolution and nastiest of fossil fuels—has just begun to lose its power over people and the atmosphere. But it is even more important to understand the underlying causes for these changes. The shift away from fossil fuels and towards renewables has been happening not because the bulk of investors suddenly became science-literate, ethical beings, but because most investors follow the money.

The easy fossil fuels—the kind you used to be able to extract with a large profit margin and relatively low risk of disaster—are essentially gone. Almost all that is left are the dregs: unconventional fossil fuels like bitumen, or untapped offshore oil reserves in very deep water or otherwise challenging environments, like the Arctic. Sure, the dregs are massive enough to keep tempting investors. There is so much unconventional oil and shale gas left underground that, if we burned it, we would warm the world by six degrees or more. But unconventional fossil fuels are very expensive and energy-intensive to extract, refine, and market. Additionally, new fossil fuel projects, at least in my part of the world, have become hair triggers for social unrest. For instance, Burnaby Mountain, near my home in British Columbia, is the site of a proposed bitumen pipeline expansion where hundreds of people have been arrested since 2015 during multiple acts of civil disobedience against new fossil fuel infrastructure. By triggering legal action and delaying the project, these protests have dented corporate profits. So return on investment for fossil fuels has been dropping.

It is no coincidence that in 2017, Petronas, a huge transnational energy corporation, withdrew their massive proposal to build liquefied natural gas infrastructure on the north coast of British Columbia. Petronas backed out not because of climate change or to protect essential rearing habitat for salmon, but to backpedal from a deal that would fail to make them richer.

Neoliberal shifts to favor renewable energies can be completely devoid of concern for climate change. While in office, former Texas governor Rick Perry questioned climate science and cheered for the oil industry, yet that did not stop him from directing his state towards an expansion of wind and solar energy. Perry saw money to be made by batting for both teams, and merely did what most neoliberal entrepreneurs would have done.

The right change for the wrong reasons brings no guarantees. Shifting investment away from fossil fuels and towards renewable energy does not mean we have entirely ditched that tired old story about fossil fuel prosperity. Once again, let's look at Perry. As US secretary of energy under Trump's presidency, in 2017 he called the global shift from fossil fuels "immoral" and said the United States was "blessed" to provide fossil fuels for the world.

In June of 2017, when I began to conceive this chapter, two articles—written by an overlapping core of authors and published in scientific journals—caught my attention. Their titles had been crafted to woo a broad audience, including journalists who might spread the word to voters and politicians: "A Roadmap for Rapid Decarbonization" and "Three Years to Safeguard Our Climate." Great shock value, I thought. Telling us that if emissions do not start declining by 2020, then it will become increasingly difficult to avoid a worsening situation. Both articles were coauthored by Hans Joachim Schellnhuber, the founding director of the Potsdam Institute for Climate Impact Research, who told the world in 2015 that "the fossil fuel industry must 'implode' to avoid climate disaster."

The studies merged climate science, international policy, and economics into a succinct plan to decarbonize ourselves by 2050, giving us a chance to accomplish the target to which world leaders agreed during the 2015 Paris Summit: to hold "the increase in the global average temperature to well below 2°C above pre-industrial levels and to pursue efforts to limit the temperature increase to 1.5°C above pre-industrial levels."

We have already reached one degree of warming, with severe impacts to ecosystems and people. Cold snaps, heat waves, droughts and floods keep breaking records, year after recent year. In 2017, wildfires in British Columbia displaced more than 45,000 people; the following year, even more. The British Columbia *Wildland Fire Management Strategy* warns that "by the middle of this century, some areas of the province may experience forest fires year-round." British Columbia spent C$380 million fighting forest fires in 2015 alone and an average of C$159 million per year over the previous 10 years.

All that seemed bad enough. Then 2018 came along, a year the BC Wildfire Service summarized as follows:

- 2,117 fires consumed 1,354,284 hectares of land, which surpassed the previously held record of hectares burned from 2017 (over 1.2 million hectare[s]).
- 66 evacuations were ordered, affecting 2,211 properties.
- The total cost of wildfire suppression reached $615 million.
- Weather was a key driver of fire activity, since late July brought about record-breaking temperatures and severe lightning storms to many areas. More than 70,000 lightning strikes lit up the province between July 31 and August 1, followed by another extreme lightning event on August 11. Within less than two weeks, the BC Wildfire Service was responding to nearly 400 new fire starts.

These facts do not even begin to capture the enormous financial costs and unmeasurable mental health damage to people forced to evacuate their homes and businesses. Meanwhile, in 2016 Canadian subsidies to the fossil fuel industry were estimated at C$3.3 billion, annually. Clearly, it makes no sense for the Canadian government to keep investing huge amounts of taxpayers' money into the root cause of expensive disasters.

The rapid decarbonization plan outlined by Schellnhuber and colleagues is the new financial narrative that, unlike most previous versions, respects the laws of physics. Fundamental to that narrative is a "carbon law" that requires us to cut global emissions in half, every decade, for the next 30 years, while simultaneously reshaping land management and our diets in ways that remove, rather than release, greenhouse gases.

Annual emissions must begin to decline—rather than merely stabilize—by 2020. No easy task. But, if we commit to doing so, and sustain the carbon law into the future, we will reach net zero emissions by 2050. Our reward will be a 75% chance of staying within two degrees of warming. Sure, as with all things scientific, uncertainty is part of this estimate. Still, the carbon law is the best shot we have.

These goals, Schellnhuber and his colleagues demonstrate, are technologically within our reach. As detailed in the rapid decarbonization plan—and corroborated by BP's Bob Dudley—we are well-poised to ramp up energy efficiency and boost the rise of renewable energies. And individuals can help get us on course by avoiding jet flights, the most climate-destroying form of travel.

Enacting the carbon law also requires us to counter the carnivorous overdrive of industrial society. Industrial agriculture currently produces one quarter of global emissions, of which 80% are caused by methane releases, deforestation, and fossil fuel use associated with cattle production. Removing industrially raised meat from our diets empowers individuals to eliminate up to one fifth of global emissions while also improving our health. Even partial vegetarianism goes a long way towards promoting social and environmental justice. And reducing food waste, at individual and more systemic levels, decreases emissions even further.

Yet, at this stage, merely reducing emissions is not good enough to enact the carbon law. Over the past 200 years, we increased greenhouse gases to levels that last occurred 15 to 20 million years ago. Those levels are not going to drop quickly on their own. So we have to reshape land management in ways that remove, rather than release, greenhouse gases. Agricultural and forestry practices that do so—some ancient, some modern—already exist and are ready to be scaled up.

Critically, Schellnhuber and colleagues argue, the carbon law is also within the grasp of our financial reality. This means scrapping fossil fuel subsidies by 2020, pricing carbon for its actual damage to the planet, and increasing government and private financing for climate mitigation projects. Global subsidies are already changing in the right direction. In 2011, they were US$523 billion for fossil fuels and US$88 billion for renewables—a six-fold difference. By 2015, fossil fuels were down to US$325 billion and renewables up to US$150 billion—a two-fold difference. Clearly, we need to keep redistributing public funds away from industries that cause climate damage and towards those that promote mitigation.

The short of it is this: full-force decarbonization must begin by 2020 or very shortly after. From then on, it must keep going. Technologically and financially, we are ready to do this.

But will we choose to do so? Will we change the stories that we tell ourselves about who we are? These questions haunt me. Which is why Jennifer Jacquet's fundamental point is worth repeating: "A geologic force is not what humanity must be." Beyond prehistoric hunters who cause extinctions, agriculturists who homogenize and simplify landscapes, and industrialists with a knack for altering the climate, there exists another way of being in the world: that of humans rooted in place and connected to their non-human kin.

Interlude II

Gail and I are camped by a lake in the territory of the Tl'etinqox band of the Tŝilhqot'in Nation, where Doreen William and Cecil Grinder will marry later that day. It is still early. A clear, crisp September morning. Some poplar leaves have begun to yellow. Goldeneyes and other ducks linger near partly submerged sedges along the shoreline. I am barely out of the tent when Cecil hunts me down and takes me to a bluff overlooking the lake so that we can "medicine up."

Doreen's father, Joseph, died the month before. The family buried him with Tŝilhqot'in prayers and no priest. Since residential schools, most funerals and weddings have been Catholic. Joseph's was the first traditional burial in a long time. After carrying Joseph's body, Cecil has yet to find time to cleanse his own spirit.

So we immerse ourselves in the sweet, light smoke of burning sage and whispered prayers.

Cecil begins to beat his hand drum and sing in Tŝilhqot'in. The song is familiar to me. Over the years we have sung it several times by the shores of Teztan Biny; I do my best to join in.

A final drumbeat. Cecil stares at the calm water before us, Datsan Chugh, where his people have always fished for trout, hunted for moose. After a long silence, he tells me, "That is how you call your people." He then explains how different sequences of beats once transmitted specific messages between hunting parties. "Something that we have forgotten." Then he corrects himself: "No. That is asleep for now."

After we return to camp, Cecil embarks on a mission. He does not explain what he is up to, yet whatever it is clearly matters to him. So he drives off in his pickup truck to a nearby campground where, the day before, he had noticed a lone RV. The behemoth vehicle is still parked there. It belongs to an older couple, two European immigrants, whom Cecil promptly invites to the wedding.

That afternoon, chiefs from two of the Tŝilhqot'in bands conduct the ceremony: the first traditional wedding i`n several generations. The event culminates in a feast that features sockeye salmon caught with dip nets in the nearby Chilko River. A resurgence of a different way of living on Earth that came into existence long before priests tried, and largely failed, to erase the diversity of this world.

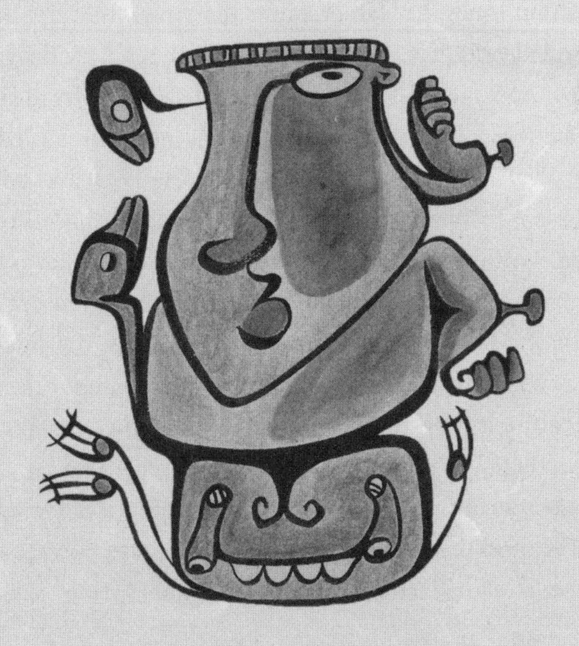

10

At the Edge of Geologic Epochs

The aluminum punt rips the glass-calm water; cloud reflections dissolve in her wake. The fog layer is high. It engulfs mountaintops, hides the sun. The air may be still, but our relative velocity transforms it into a 25-knot wind against our bodies, and we have no shelter. Eleven-year-old Kvai-Lynn—pulled out of her warm bed much too early, stoic against the chill—slumps low on her folding lawn chair, temporarily hides from our world inside the refuge of her hoodie. Colin Jones, her father, pilots the punt while sipping coffee from a lidded mug. He is all smiles, as always. The village of Bella Bella, the home of Kvai-Lynn and Colin, disappears from sight.

It is late May of 2018. We are out to harvest the seaweed known as *łq'st* to Kvai-Lynn's and Colin's people, the Heiltsuk, and as the genus *Pyropia* to Western science. Throughout the coasts of British Columbia, Washington, and Southeast Alaska, *Pyropia* is such an important food and medicine to First Nations that there are as many names for it as there are Indigenous languages. Through trade, *Pyropia* also is known to interior peoples; the Tŝilhqot'in, who live inland, acquire it from their coastal neighbors, the Nuxalk, and call it *tish-guns*, which translates as "underwater scale-like."

Pyropia is an annual. It renews, grows, reproduces, senesces, and dies each year on the mid-intertidal zone of rocky shores. Colin and family have already been out several times this spring, drying their seaweed back home on racks and rooftops, where the spring sun concentrates its heat. Today we are heading to Chugways, where we will harvest until the rising tide reclaims the shoreline. We are targeting the second growth of the season, which some consider to taste even better than the first.

Traditional knowledge says the harvest stimulates growth; Helen Clifton, from the Gitga'at Nation of British Columbia's North Coast, described the harvest of *ła'ask* (as *Pyropia* is known in her Coast Tsimshian language) to the ethnobotanist Nancy Turner: "It's just like any garden, it has to be tended. So if you pick it every year, then it grows strong the next year; it keeps coming back. So if it isn't picked for a few years, then it just has rotted away on the rocks there."

Whenever I participate in harvesting trips, I am struck by how they are about so much more than food. As we travel on, Kvai-Lynn pokes her head out of her hoodie to observe sea otters, harbor porpoises, and the tiny seabirds known as marbled murrelets. The moment we land at Chugways, she is off to the races. The low tide has exposed a field of *łq'st*, and she energetically fills large bucket after bucket.

This is her upbringing, yet I worry about how many other children will be raised harvesting *łq'st* in the future.

The seaweeds along the shores of the northeast Pacific Ocean are locally adapted to a narrow range of ocean temperatures. A slight warming above that range can slow down their growth and productivity. During The Blob years of 2015 and 2016, lighthouse keepers at McInnes Island, in the Central Coast, recorded average sea surface temperatures of nearly 11°C during April—almost 3° warmer than the average for that month during the preceding 60 years. Seaweed harvests crashed, which filled people with fear: for their food and culture, for their trading relationships with neighbors, for kids like Kvai-Lynn.

How long will the reprieve last?

Sea surface temperatures keep rising.

Blob-like conditions will spike higher.

With *łq'st* brimming out of buckets, we hop back in the punt. The tide is not too high yet, so we slowly motor to a nearby offshore rock and grab a half-dozen red urchins from its sidewalls. We crack a couple open and snack on their bright yellow gonads. Kvai-Lynn is visibly delighted to be eating raw innards of the freshly killed urchins straight out of the shell. Just as we are about to accelerate and start heading back to town, we watch a wolf lope along the *łq'st*-covered shore.

Back in Bella Bella, a festive atmosphere is palpable. Just about everyone has *łq'st* drying on racks in their yards and on their roofs. Upbeat greetings slow my walk through the streets.

A wonderful change from the scare of The Blob.

The Nuxalk recall at least three times when *sputc* (eulachon) disappeared from the Bella Coola River and had to be brought back from other rivers, twice by the people, once by Raven. The Cree of northeastern Canada have known caribou to "disappear under the water," sometimes for decades, each time expecting "a time when they would reappear later." These events precede industrial civilization. Yet the scale of these resource crises experienced by First Peoples in the past is so much smaller than our escalating rate of global change.

There are 27,600 known species of vertebrates on Earth today. One third are declining in abundance and geographic distribution. In the past century, nearly 200 species of vertebrates have gone extinct. "Biological annihilation." That is the term ecologists use to describe what industrial civilization is doing to our non-human kin.

A ball inside a cup. That is one metaphor ecologists use to describe resilience. The cup is the range of shocks—climate change, logging, fire, industrial fisheries—that the ecosystem can withstand without losing its essential relationships, the interplay between organisms and energy flows that characterize that ecosystem. The ball—which rolls up and down the sides of the cup in response to shocks of different magnitude—is the state of the ecosystem at a given moment. The stronger the shock, the closer the ball nears the edge.

Resilient ecosystems have no idealized, fixed place at the bottom of the cup, where stability is greatest. They keep changing yet maintain their essence while the ball rolls high or low. New species show up—on their own or introduced by people. Old species leave or go extinct—on their own or exterminated by people. Both processes change the identity and number of ecological actors. Still, the ball remains within the confines of the cup as long as the main roles played by ecological actors stay roughly the same. But if the combined shocks become too much, ecological roles fall apart, and the ball flies off into a different cup, the cup of an alternative stable state from which return to the original is unlikely.

I worry that our studies of yelloweye rockfish might be signaling the early stages of an alternative stable state. Through a combination of fisher knowledge and scientific surveys, our research suggests that the average adult yelloweye today is nearly one half the length of an average adult yelloweye 30 years ago. The relationship between weight and length in many fish species is exponential, which means that in becoming shorter, yelloweye also have become disproportionately thinner and lighter. Frank Johnson from the Wuikinuxv Nation recalls that, until the 1980s, yelloweye rockfish weighing 7–9 kilograms were a common catch for his longline sets at depths of 50 to 100 meters in inside waters. These days, at those same depths and places, the yelloweye Frank hauls weigh a one third to one quarter of his past catches. And yelloweye, along with other rockfishes, continue shrinking. Because their fecundity increases with size, the trend toward smaller fish signals tremendous drops in the productivity of populations that already are deteriorating.

Commercial and recreational fishers initiated the shrinking of yelloweye and other rockfishes by targeting the largest fish. They are still catching whatever big ones are left. Yet even if we halt those fisheries altogether, it may no longer be possible to restore the larger fish sizes that Indigenous elders caught regularly in the past. Rising sea temperatures and lower oxygen levels in the ocean are symptoms of climate change. They lower the metabolic rates of fish, curtailing their potential to grow to their former maximum sizes, which has implications that extend beyond individual fish species.

At least historically, yelloweye can grow to nearly one meter in length. Fish that size are potentially important predators of other fish and crustaceans, which means that they might influence the distribution and feeding behavior of different kinds of prey, perhaps playing important roles in structuring rocky reef communities. Wherever large yelloweye have disappeared, so have their interactions with larger-sized prey, which may trigger cascading effects into other parts of the food web. A large body of research tells us that when the largest predators are gone, biological communities simplify, losing their species diversity and resilience. They undergo what the ecologist John Terborgh calls "transitions to radically distinct alternative states."

And the relationship between rockfishes and people also is changing. Mike Reid from the Heiltsuk Nation recently wrote to me:

> There has been a huge decline in [the effort of] our fishers fishing cod [rockfish] species using jigging lines over the last 20 years, and from personal experience I know it's related to the decline in rockfish populations in areas where we used to get rockfish for food, social, and ceremonial purposes. I recall bringing my family out to go jigging and not having any success at all in areas we knew had rockfish in the past. Now, most of the effort is directed at getting halibut [with longlines] and you get a few rockfish during that process. I don't

recall seeing any of our folks jigging for the longest time now, 8–20 years, and jigging used to be a common practice. This partly accounts for low catch data reports, folks are not targeting rockfish anymore.

Are yelloweye rockfish permanently dwarfed—in size, ecological function, and contribution to culture? What does this mean for the resilience of rocky reef communities? For the resilience of the people?

The answer to these questions partly depends on the status of other large predatory fishes that might play ecological and cultural roles resembling those of yelloweye rockfish. Lingcod also live on rocky reefs and are important as traditional food. They grow even larger than yelloweye and certainly are voracious eaters of fish and crustaceans; during our studies we have found adult pink salmon and copper rockfish inside the stomachs of meter-long lingcod. Lingcod have a much faster life history than rockfishes. With a maximum known life-span of only 25 years, they grow quickly and start reproducing when only three years old, which makes them less vulnerable to overfishing. Lingcod rarely venture into the 150- to 400-meter depths that yelloweye and other large, long-lived rockfishes prefer. So my guess is that species like lingcod will buy us some time in the shallower parts of the ecosystem, contributing to ecological and cultural resilience further into the future than rockfishes. But how long can predators like lingcod hold down the fort in the face of climate change and fisheries? That I do not know, and the question troubles me.

There was a moment, about 10 years ago—while writing letters to be read in the future by my daughter, who was five at the time—when I accepted that there is no going back to the planet onto which I was born, much less the one elders alive today knew during their youth.

It sobered my expectations for how much we could restore degraded ecosystems to their former states, no matter how well we might suddenly start managing ourselves and conserving our resources.

The years between 1945 and 1964 are increasingly recognized as the end of the Holocene and the start of the Anthropocene. Granted, scientists of the Anthropocene Working Group of the International Commission on Stratigraphy have yet to formally acknowledge the onset of a new geologic epoch, but they are close. In their own words, when they do make it official, "not only would this represent the first instance of a new epoch having been witnessed firsthand by advanced human societies, it would be one stemming from the consequences of their own doing."

Sometimes I have to fend off the temptation to fall off the wagon. To indulge again in giving up, which is so much more comfortable—anesthetizing, really—than working towards what humans *can* do.

Imagine Indigenous children growing up in the thick of the era of institutionalized racism—which in Canada has just begun to end—when federal laws criminalized their language, ceremony, and tradition. Now picture these same individuals as elders today, instigating a cascade of young people eager to rediscover traditional place names, practices, stories, laws. Reviving landscapes and seascapes with the breath of living cultures.

I have been witnessing some of this momentum while helping design in British Columbia a network of marine reserves in which the biological and cultural areas designated for protection are identified by combining traditional knowledge and science. The process has included First Nations as co-governance partners with the federal and provincial governments. This is a big deal, a recognition that people lived in and managed their territories long before Europeans arrived. By becoming a level of government, Indigenous peoples can regain control over their resources in ways that safeguard biodiversity and their cultural practices while still being largely (though not completely) inclusive of the general public. The network of marine protected areas is not a done deal yet. It may not be for another year or so, and it can be squashed—or at least significantly diluted and diminished—by political short-termism. Yet the fact that the process exists, and appears to be genuine, gives us something we can work with.

Hereditary chiefs, as responsible stewards of specific places, traditionally can close areas to any kind of exploitation whenever deemed necessary by their knowledge and that of their advisers. This mandate has always been important, yet it is arguably even more critical today, in the Anthropocene. Climate change or other environmental shifts do not affect biological communities in isolation. Rather, they combine with other stressors—including fishery exploitation—in ways that can be stronger than the sum of their parts. The First Nations I work with often invoke their hereditary chief system when working with the federal and provincial government towards the establishment of marine protected areas. When they do, cultural resilience and ecological resilience can become one.

A warming ocean is stressing marine organisms, yet marine reserves that exclude fisheries can give those organisms a better chance of withstanding environmental change and, in the process, give Indigenous people a better chance of maintaining access to traditional resources. Sure, as the ocean warms and acidifies, some species will vanish, novel species will replace them, and some ecosystems *will* shift to alternative stable states. We will never go back to the abundance that existed in the living memory of Indigenous elders. But we can steer things in ways likely to keep the basic components—the top predators, the prey, the plankton, even if some are replaced by novel species—still interacting with each other.

And with people.

Day after day, Gail, Twyla Bella, and I walk with our friends Pierre Friele and his son Kai on fields of obsidian and among volcanic cones, high in the mountains of northern British Columbia, where the Anthropocene interfaces with earlier geologic epochs. The obsidian exploded out of the earth during the volcanic eruptions of the Miocene, perhaps seven and a half million years ago, yet has been continuously churned to the surface by freeze–thaw processes and glacial dynamics for the past 11,600 years, since the start of the Holocene. The Tahltan and their neighbors began shaping obsidian into blades and projectile points early in the Holocene and continued to do so until colonization disrupted their way of life.

We pass dozens of Stone's sheep resting on boulder fields near the edge of Mount Edziza's glaciers. Spook a woodland caribou mother with her calf. Intersect the tracks of wolves, moose, and grizzly bears at the bottlenecks of mountain passes.

Arctic ground squirrels stand erect by their burrows; miraging heat waves amplify these rodents into lighthouses.

The white coat of a lone mountain goat contrasts against red pinnacles of volcanic rock.

Willow ptarmigans burst out of shrubs like mottled, feathered cannonballs on low arc trajectories. Their mid-air calls explode into an accelerating blend of barks and drumbeats that, upon landing, slow down into hoarse, cartoon-like coughs.

On a high plateau and near the edge of a ravine, overlooking a valley, Pierre, a geomorphologist who collaborates with archaeologists, points to chipped pieces of obsidian on the ground. He remarks, "Someone sat here, making arrow points and looking out for game below." It is shortly after that I spot, through binoculars, a wolverine scampering across the lower valley.

Hauling ourselves up and down the steep screes and shale-covered slopes of the Spectrum Range we are often hungry. Into our meals I put dry herring eggs I collected in Kitasu Bay with Charlie Mason and dry seaweed I collected with Kvai-Lynn and Colin at Chugways.

On a ridge top I examine a large chunk of obsidian: black, smooth, shiny. On its bulb of percussion—the wave pattern that emanates from the epicenter where a long-ago hunter struck stone against stone to initiate the making of a blade—I see my reflection.

Throughout our journey, the merged Alkali Lake and South Stikine River fires burn nearby, massively, eventually consuming 1,212 square kilometers of forest and most of the community of Telegraph Creek: an expression of exceptionally hot and dry conditions at 57°N. Pyrocumulonimbus clouds generated by the flames chase after us, draw near, yet do not catch up.

11

Transformation

The ancient teacher is some thirty meters below the surface. I never know when I will be granted an audience, but whenever it is, it always sears itself into my memory, shifts my perspective, jostles me out of wherever I might be stuck and into something much bigger than myself.

So I jump off the boat and immediately struggle to keep the strong tidal current from sweeping me away.

Near the bottom, that current subsides.

Clear water.

Floating in stillness.

A mere six kilometers from my home as the raven flies, this is where nearly a decade ago I began to realize that nihilism was not going to get me anywhere I wanted to be. That the Anthropocene has a range of possibilities that have yet to be realized, and that we humans still have agency to use our proven capacity to conserve and aim for the better part of that range.

The rocky reef gives way to a bioherm: a vast expanse of cloud sponges growing on layer upon layer of dead cloud sponges that have been accumulating for millennia. I hover above the living ones: convoluted, amorphous yet somewhat tubular, yellow-and-white beings built of silica crystals and soft tissues. The bioherm is the submarine equivalent of an old-growth forest. Like trees removing atmospheric carbon and sequestering it into wood, the sponges are filter feeders that consume vast amounts of bacteria. Each square meter of bioherm sequesters up to one gram of bacterial carbon into its tissues every day, reducing the rate at which our greenhouse gas emissions acidify the ocean.

Several quillback and redstripe rockfish and a juvenile yelloweye rockfish rest among the sponges. I take in the fishes' deeper oranges, stronger yellows, and darker hues that contrast against the sponges.

Until I enter a school of at least 500 yellowtail rockfish circling above the bioherm. Drawn into that sweeping, living energy, I am no longer its witness.

I become it.

Inseparable.

It is at times like that this that I truly understand what it means for gravity to be suspended. That there is no choice but to do what we must.

A week before the bioherm dive, on November 23 of 2018, Mike Reid wrote me an e-mail that began, "I have often found in our work with DFO and others that our knowledge is seen as trivial and not taken seriously." Yet, despite that gloomy opening, Mike was writing to express that he felt that things could really be changing. He continued:

> At our last meeting with DFO on herring, Colin Masson [a senior manager] shared that he didn't believe us last year when we were pushing for no commercial sac roe fishery; he implied that all he saw were the numbers from the model telling him that there would be lots of herring. He apologized.
>
> Traditional knowledge has different parts to make a whole, teachings that are precautionary, teachings that say to give back, and the biggest teachings are in the long history of observations passed from fathers and mother to sons and daughters.
>
> I see that science is catching up in some areas, and one in particular strikes me more. There is research in which carbon

> was treated with a nuclear tracer and then given to a large tree; they used big plastic bags over the branches and the tree sequestered the carbon, and they traced the carbon going in the tree, going all the way up and then coming down and going out to a very small tree nearby. The big tree was feeding the small tree. They now find that if you leave large trees in a cut block, the trees around grow faster and healthier.

Mike was referring to the brilliant experiments by forest ecologist Suzanne Simard and her colleagues. I then asked Mike whether his people already knew what the outcomes of the experiment would be.

"Of course," he replied.

I do not advocate a switch to a global economy in which we all make a living by hunting, gathering, fishing, and tending native species. If we could, we would be healthier, with more time in our hands and less clutter in our minds; we would be better connected to our relatives and the beauty of this world. But we are hundreds of years too late for that.

And that is just as well. Modern civilization—its crazy and wonderful melting pot of ideas, songs, dances, and beauty—captures some of the best that humans can offer. But that is just icing on the cake. The real reason why I would not revert to a past era is that modern civilization has been, rapidly and steadily, eradicating the violence that has permeated almost all human cultures throughout history.

The ancestors of my Indigenous friends and colleagues may have trodden gently on Earth but not necessarily on each other. Tribalism, slave raids, warfare, social and gender inequalities: these were part of their ways of life, though no more so than they have been to most other cultures in the past.

What really matters is that different First Nations who warred against each other historically are at peace today—collaborating towards a revitalization of their stewardship practices. When I participate in marine-planning working groups, I take it for granted that members of different First Nations now sit at the same table, working together. Instead of fighting, as their ancestors might have done, they now gather around campfires to showcase their traditional songs in friendly competitions. And, just for old times' sake, everyone still gets to duke it out during inter-Nation basketball tournaments.

Similar patterns scale up globally. Despite the ongoing damage of religious fanatics, US gun culture, Donald Trump, and other forms of extremism, modern civilization is less violent than its predecessors. As the scholar Steven Pinker points out, since the end of World War II global violence has declined steadily. Sure, imperialism is not gone, and in many parts of the world, including sections of some Canadian cities, the wrong skin color remains akin to a death sentence, often executed by the police. Still, the death toll of imperialism and bigotry is declining and becoming a harder sell, at a societal level, by those who wish to perpetrate it.

Which means that we have a tremendous opportunity to keep doing better than our ancestors, both distant and recent. We can ditch the destructive stories that began when we first burned coal in the steam engines of the Industrial Revolution. We can merge the best of the past and present into something better than what we currently have, than we ever had before. To seize this opportunity we need to embrace many scales of transformation, from the personal to the societal.

And human transformations can be story-driven.

"There is no fixed 'human nature,'" wrote the ecologist Fikret Berkes and the ethnobotanist Nancy Turner. That simple statement conveys a profound message. Humans are not destined to be climate-altering, species-annihilating, self-destruction machines. Rather, we are destined to be the story that we tell ourselves of who we are.

Sure, humans are subject to our evolutionary history and impulses, just like other animals. And like other organisms, our genes want us to consume resources selfishly so that we may leave as many descendants as possible in the world. If human actions, in concert with our powerful technologies, were determined by nothing but that drive, biological annihilation would be inevitable. Yet, as Berkes and Turner argue, "conservation practices can develop over time" because common property institutions can temper our impulse to relentlessly consume resources and maximize reproductive success.

I spent the earlier part of my career expanding and testing "optimal foraging theory," which posits that organisms maximize their lifetime reproductive success by making efficient feeding decisions while simultaneously avoiding becoming a meal for somebody else. "Central place foraging" is a subset of that theory that predicts and interprets how animals that return regularly to a "central place," such as a burrow or a nest, decide where to feed and how much time and effort to spend feeding at a resource patch. For example, nesting birds travel to find food and carry that food back to their chicks. To produce as many viable young as possible over their lifetime, the time and energy they invest in a round trip must be worth the expected food reward. That is, when good food is nearby, it is wasteful to travel far and better to do many short trips. But as nearby food becomes more scarce, nesting birds and other central place foragers, such as mammals that carry food back to their young in a den or burrow, go farther afield, often making fewer trips (because each trip requires more time) and bringing back more food per trip (because a bigger reward compensates for more travel). Animals in the real world do behave that way, for the most part.

When central places have a high density of animals, as would be the case with a colony of thousands of seabirds, we often see a halo of depleted resources near that colony. That is, the seabirds eat the nearby and larger fish first and, as the closer and larger pickings vanish, the birds keep expanding the halo of depletion farther from their colony and switch to consuming smaller prey. That situation is analogous to what might happen around villages or other human infrastructures. Central place foraging, therefore, can help us test for situations in which people may have deviated from evolutionary drives and chosen to conserve, or not.

In one project, I used the central place framework to infer the footprint of sport fishers on rockfish. These fishers often operate from small boats and use harbors or other coastal infrastructure to repeatedly start and end their daily trips. I hypothesized that the cumulative number of trips to a site might decrease as the fuel and time costs of travel increased with distance from the harbor. That line of thinking led me to predict that the impact of sport fishers on stocks would be greatest near lodges or similar central places and weaken with distance. Consistent with these ideas, field data I had collected with Central Coast First Nations and analyses I had conducted with my colleague, Madeleine McGreer, showed that the greater the distance to a central place, the larger the size of yelloweye and quillback rockfishes, and the greater the collective abundance of long-lived species of rockfish. That finding suggests that, within the confines of regulations that limit their gear types and catches, sport fishers behave as optimal forages. Of course, this is to be expected. Most recreational fishers do not live near the places they fish, so they would be hard-pressed to skip over a good fishing spot near port with the intention of safeguarding fish there for the future. As legal roving bandits, they are doing what is best for them during their holiday.

I am certainly not the first to apply foraging theory to fisheries or other forms of human behavior. Some anthropologists have used this framework to infer cases in which Indigenous peoples also behaved as optimal foragers and not as intentional conservationists: hunting animals as they encountered them—not selectively, according to a prey's gender, size or greater abundance—and depleting resources near their villages in an expanding halo. But the fact that these cases have occurred—indicating situations in which sustainability requires human populations to remain small and sparse—does not imply that they are the norm. And they certainly are not representative of First Peoples in coastal British Columbia.

Rockfish bones from middens at five villages in Haida Gwaii, a large archipelago located 100 kilometers off the British Columbia mainland, illustrate this point. Archaeologists analyzing those bones for stable isotopes, biomarkers that provide a chemical signature of where rockfishes spent their lives prior to being caught, inferred that the people who produced those middens generally harvested rockfishes near home. At first this may seem like evidence of central place foraging: you harvest what is nearest first. Yet there appeared to be no evidence that these pre-colonial fishers created an expanding halo of depletion. People fished for rockfish near their villages for periods that ranged from 400 to over 1,500 years, until European colonization ended their way of life.

Anyone with access to basic fishing gear and a small boat who is not practicing intentional conservation *will* quickly deplete rockfish at specific reefs. If Haida fishers had lacked stewardship practices, then rockfish depletion would have occurred centuries before European contact. In that case, only the earliest bones would have come from fish harvested near the villages, while younger and younger layers of bone would have come from farther and farther away. Those patterns would have been consistent with a depletion halo expanding over time, and a smoking gun for optimal foraging over intentional conservation. But that is not what evidence supports.

I am not implying that Indigenous "foragers" never optimize. There is no doubt in my mind that human hunters and resource gathers optimize to some extent—I do it myself, and I certainly have watched my Indigenous friends do so. But to leave it at that would be simplistic. People are much more than optimization algorithms. As Berkes and Turner have pointed out, "The optimum foraging approach to conservation is blind to the fact that people and societies are capable of learning from experience, modifying their decisions and rule sets, and passing their knowledge on to others."

And this is where connection to place is key. Roving bandits, legal or otherwise, will always optimize because what they leave behind, or what they do not, is someone else's problem. When scarcity hits, they simply move on and will continue to do so until biological annihilation is a fait accompli for the whole planet. Yet when people, whether Indigenous or naturalized, become inherent to place, cultural conservation ethics can supersede evolutionary drives.

Seventy-year-old Eleanor Schooner, a Nuxalkmc whose traditional name is Hrwana, sits in her fish-processing shed, filleting a big load of chum salmon her relatives just caught in the Bella Coola River. Torrential rain pounds the tin roof: the very storm that is about to dislodge a landslide into the river, muddying the waters and slowing down catches for fishers using gill nets.

Hrwana tells me how her parents lived by the Kimsquit River, near the head of the fjord now known as Dean Channel. As hunters of deer and mountain goat, they would pray to Alquntam, the Creator, for the amount of game that they needed. Alquntam always provided, and her parents never took more than what they had asked for. In 1938, ten years before Hrwana's birth, her parents were the last to leave Kimsquit and move to Bella Coola, where the Nuxalkmc, whose numbers had declined because of disease and other colonial impacts, could regroup and support each other as a community.

In Bella Coola, traditional practices prevail. Hereditary chiefs close the river to salmon fishing during weekends. Hrwana speaks of how, in her younger years, harvesting eulachon for grease required a steady month of work. "If you do not get that oil, then you are in trouble," she says. Yet, as a precautionary measure, her family harvested eulachon only every second year.

Eulachon had been absent from the Bella Coola River for over 20 years when, suddenly, they showed up earlier that spring. Yet the people did not harvest them; the eulachon, too, needed to regroup and find their strength first.

Do traditional stewardship practices evolve preemptively, or are they a learned response to catastrophe? Turner and Berkes argue that both processes occur. The Cree of northeastern Canada first acquired repeating rifles in 1910 and were quick to deplete the caribou herds that, up until that point, they had hunted sustainably. The event stunned the Cree, who had underestimated how powerful new technology could cloud their vision. But they recognized their mistake and adapted their conservation protocols to harness the greater destructive potential of new tools.

This is one example of how gentler feedbacks between actions and reactions prevailed in the way humans, equipped with new, more efficient tools, learn to gauge and adjust their relationships with ecosystems. In the age of technologies that can be far more destructive than the repeating rifles of 100 years ago, the challenge to continue to do so is ever more difficult, yet not impossible to meet.

The transition from conflict to cooperation between many First Nations was a relatively quick response to an externally imposed catastrophe. Colonial diseases and policies, which in many ways are one and the same, disrupted all forms of governance, social cohesion and transmission of intergenerational knowledge. Within a few short decades, they weakened powerful, rooted cultures, blasting them out of their homes and leaving them wounded.

But not dead.

As First Nations continue to heal—to re-empower themselves, to rebuild clam gardens, to apply traditional laws to shut down commercial fisheries, to revive stories that speak of responsibility to all living things—they chose to do so while leaving behind, for the most part, the tribalism of old. Transformation, both on the individual and societal levels, has followed the shock, disruption, and systemic horror of colonization. Resurgence into something more peaceful than before, without losing sight of the philosophy of kinship with our non-human relatives that the entire world needs if it is to steer towards a more benign Anthropocene.

Catastrophe can be our teacher, but it need not be the only one.

Almost always, I stay away from dualistic perspectives. Good and evil. Heaven and hell. Us and them. I find these dichotomies limiting. I relate more to the blurred edges, gradients, and non-linearities that characterize many traditional Indigenous world views, which differ only so much from how ecologists see the natural world. Yet Indigenous stories go deep into territories where ecologists cannot venture, invoking the spiritual and supernatural. As an empiricist, I interpret these as metaphor.

Physicists have summarized all known physical laws into the beautiful sinuous symbols of a single equation: precise, elegant, yet only a map to the universe, not the universe itself. A topographic map is a two-dimensional image that represents vast landscapes with intricate corrugations of topography and vegetation variants: incredibly useful, mostly accurate, yet only metaphor.

Stories are the same. They just happen to pick up where equations and topographic maps cannot go. Believing, as I do, that everything can be explained in terms of the observable and the measurable—including the chemicals that make me experience love, the rush of running through the forest, the sense of grounding I derive from writing these words—does not preclude me from feeling how stories give meaning to our lives.

It is the stories I connect with that make me get up in the morning, hug my daughter, write this book, cry over intergenerational justice, tell my wife that I love her more than I can say, feel rage over the racism that has pushed some of my friends into alcohol and self-destruction...be carried by the thunder of Cecil's drum as it echoes across the still waters of Teztan Biny. They keep me feeling and believing, beyond any doubt, that these are the collective reasons why I, we, exist at all.

The next many millennia on Earth will reflect the stories that we the *Homo sapiens*, the tool inventors and users, decide to ditch or accept into our collective identity during the second and third decades of the 21st century. We are on a knife edge, and this is the time to act.

TAPOUT

BROOKS

Acknowledgments

I am deeply grateful to the Wuikinuxv, Nuxalk, Heiltsuk, and Kitasoo/Xai'xais peoples, who gave me the best possible job I could have wished for—one that has allowed me to gain many of the insights I describe in this book—and to the Tŝilhqot'in people, for all the experiences they have allowed me to be part of. Individuals from these First Nations have given me tremendous gifts of generosity, wisdom, knowledge, humor, ocean and forest food, ceremony, song, art, friendship, and most other things worth living for.

To my colleagues from the Central Coast Indigenous Resource Alliance, University of Victoria, Simon Fraser University, Florida State University, and Fisheries and Oceans Canada, I also express my gratitude.

The book originated in the late summer of 2017, during forest walks with my wife, Gail Lotenberg. Words will always come short in trying to express my love and gratitude to Gail, for her insights and connection during those walks and for everything else that pervades our lives together.

Once convinced that the book was worth pursuing, I turned to my friend, writer James B. Mackinnon, who reviewed the earliest drafts and set me on a forward path. Later on, Neasmuutk Haimas (Charlie Mason), 'Cúagilákv (Jess Housty), Cecil Grinder, Nang Jingwas (Russ Jones), Kanilkas (Alvina Johnson), Fred Smith, Kyle Artelle, Nancy Turner, Larry Dill, Charlotte Whitney, and Iain McKechnie commented on individual chapters, and Q̓án̓ístisḷa (Michael Vegh) reviewed the entire manuscript. Cecil and Jess also provided timely feedback on the cover. My friend Pierre Friele helped me interpret archaeological and geological studies, and supplied me with relevant literature. To all of you: gratitude for your time, knowledge and critical thinking.

While gearing up towards more advanced stages of the book, I had the great fortune of being introduced by my sister, Dianna Frid, to Anna Badkhen. Anna, a superb thinker and writer with a huge track record of books and journalism, became my main editor, working me hard each time my writing, logic, or awareness of sensitive issues slacked below its potential. Thanks to her, the book is stronger; any remaining weaknesses are my own. I also thank Anna for capturing the geography of the book into a beautiful hand-drawn map.

I am grateful to Tristan Blaine, Adam Taylor, Doug Neasloss, Markus Thompson and Pierre Friele, who donated use of their fantastic photographs conveying research or life in the Central Coast (TB, DN, MT), dives at the Halkett sponge bioherm of Howe Sound (AT), or the pyrocumulonimbus cloud at the edge of the Edziza plateau (PF). I also thank Gail for her essential advice during image selection.

Funny how the trials and tribulations of undergoing dental surgery can sometimes send you down a lucky path. On the afternoon of May 31, 2019, when I thought I was done writing and editing for this book, I waited for the ferry, trying to get back home. I had spent the morning listening to drills reverberate inside my cranium and getting a tooth yanked out. I longed for a dark room to crawl into, not conversation. That is when Michael Nicoll Yahgulanaas showed up. Michael and I live on the same smallish island. We see each other only once or twice a year, but it is always a great connection. So his sudden appearance banished my sulking, and we got talking. Within moments of me mentioning the "new" book, Michael offered to contribute his artwork. What ensued was a last-minute and extremely rewarding collaboration that led to a revised preface and to Michael's images presiding over the start of each chapter. Michael, a big *Háw'aa* to you.

The physical book would not exist without New Society Publishers (NSP), who gave me the means to connect with readers. While I am indebted to the whole NSP team, I am particularly grateful to Rob West (acquisitions editor) and Betsy Nuse (copy editor).

During the year in which I did the bulk of the writing, I witnessed our daughter, Twyla Bella Frid Lotenberg, begin to come of age: becoming someone who loves Earth (and black holes) for her own reasons and initiating a quest for her own story. Twyla Bella gives me a reason to keep writing; it is my hope that this book will do its small part in inspiring her—and the rest of us—to do all we can to rebuild a world where people from different cultures relate to each other, and to our non-human kin, with respect, reciprocity and love.

Captions

viii Neasmuutk Haimas (Charlie Mason) and his grandson, Dean Duncan, from the Kitasoo/Xai'xais Nation, amidst the Kitasu Bay landscape, March 2015.

xiv Kitasoo/Xai'xais youth in traditional regalia shortly before a welcoming ceremony for canoeists arriving at Klemtu from First Nation communities in Haida Gwaii and the Skeena River region, July 2014

xviii A school of subadult widow rockfish observed during a dive survey near Calvert Island, October 2018. Credit: Tristan Blaine.

12 Heiltsuk hereditary chiefs in traditional regalia welcome paddlers to Bella Bella during the Qatuwas festival, which celebrates the resurgence of Indigenous canoes, July 2014.

26 Ernest Tallio (Nuxalk Nation) maneuvers a rowboat while using a drift net to fish for coho salmon on the Bella Coola River, October 2015.

48 Kitasoo/Xai'xais researchers Sandie Hankewich and Ernie Mason survey Dungeness crab at Mussel Inlet. Credit: Tristan Blaine

58 Elmer Starr gathers his share from a harvest of herring eggs on hemlock boughs, Klemtu, March 2017.

78 Markus Thompson during a research dive in Spiller Channel, April 2016. He is examining the millions of herring eggs that cover tubeworms attached to a submerged cliff, continuously, from the surface to a depth of about 40 meters. Credit: Tristan Blaine.

92 Grizzly bears eating salmon at the Mussel River. Credit: Doug Neasloss.

105 George Johnson describes his work in progress—a depiction of the relationship between salmon, bears, killer whales, eagles and Wuikinuxv people—which he has been carving into a cedar log, Wuikinuxv Village, April 2019.

119 Tŝilhqot'in children at Teztan Biny, October 2017.

136 A pyrocumulonimbus cloud generated by the merged Alkali Lake and South Stikine River fires of 2018, as seen from the Edziza Plateau of northwestern British Columbia. Credit: Pierre Friele.

138 I examine a juvenile yelloweye rockfish (*center*) at the Halkett sponge bioherm of Howe Sound, December 2018. Also visible are a quillback rockfish (*right*), yellowtail rockfish (*far right*) and a redstripe rockfish (*left*). Credit: Adam Taylor

150 People from different bands of the Tŝilhqot'in Nation gather for an annual ceremony at Teztan Biny.

Color Section (plate numbers)

1. Gilbert Solomon (Xeni Gwet'in band of the Tŝilhqot'in Nation) sings during a ceremony by the pole carved by Nuu-Chah-Nulth artist Tim Paul. Teztan Biny, October 2017.
2. Heiltsuk people in traditional regalia welcome paddlers to Bella Bella during the Qatuwas festival, July 2014.
3. Robert Duncan (Kitasoo/Xai'xais) travels on Charlie Mason's punt during herring season in Kitasu Bay, March 2015.
4. Kitasoo/Xai'xais children, Klemtu, May 2016.
5. Ernie Mason (Kitasoo/Xai'xais) emerging from a research dive in Kitasu Bay, August 2013.
6. Yelloweye rockfish observed during one of our dive surveys in the Central Coast; a quillback rockfish is partly visible inside the crevice. Credit: Tristan Blaine.
7. Quillback rockfish observed during one of our dive surveys in the Central Coast. Credit: Tristan Blaine.
8. Leather star on herring-egg-covered reef, Kitasu Bay, March 2016. Credit: Tristan Blaine.
9. I examine a school of yellowtail rockfish at the Halkett sponge bioherm of Howe Sound, southern British Columbia, December 2018. Credit: Adam Taylor.
10. Norman William (Xeni Gwet'in) pauses during a deer hunt in the Nabaŝ area, near Teztan Biny, October 2017. He stands by the remains of the remote cabin where he lived with his family for much of his youth.
11. A spirit bear cub carries a salmon carcass by one of the creeks that drain into Kitasu Bay. Spirit bears are a rare subspecies of black bear in which 10 to 20% of individuals have white pelage. Parents with black coats can produce white offspring. Credit: Doug Neasloss.
12. Nuxalk Sputc (eulachon) pole by the Bella Coola River. The pole—created by Lyle Mack (Wiiaqa7ay) with contributions from several other Nuxalk artists and raised in March of 2014—represents Raven (Qwaxw)

carrying a male and female Sputc. This pole is emblematic of the teachings that many Nuxalkmc continue to live by.

13. The colors of Kitasu Bay transformed by milt released by spawning herring, March 2016.
14. Surf scoters in Kitasu Bay during herring season, March 2016. Credit: Markus Thompson.
15. A school of yellowtail rockfish feeds on a school of Pacific herring, as observed during a research dive in Draney Narrows, Rivers Inlet. The stomachs of the rockfish are visibly full. Credit: Tristan Blaine

Notes

Chapter 1: Gravity Suspended

My work as an ecologist is with the Central Coast Indigenous Resource Alliance (ccira.ca), which comprises the Wuikinuxv, Nuxalk, Heiltsuk, and Kitasoo/Xai'xais First Nations of coastal British Columbia, Canada. For an outcome of the dive research described in this chapter, see Frid et al. (2018). For an introduction to the natural and life histories of rockfish, see Love et al. (2002). To learn about longevity overfishing and its implications to fisheries sustainability, see Berkeley et al. (2004), Beamish et al. (2006), Hixon et al. (2014), and Barnett et al. (2017). For a synthesis on the relationship between body size and fecundity in rockfish, see Dick et al. (2017).

Evidence for 14,000 years of human occupation in the Central Coast of British Columbia comes from recent archeological excavations at Triquet Island (MacLaurin 2017). Evidence for 9,100 years of rockfish fisheries in coastal British Columbia comes from archaeological research in Richardson Island, Haida Gwaii (Mackie et al. 2011). McKechnie & Moss (2016) quantify the ubiquity and relative abundance of fish bones (including rockfish and halibut) in archaeological sites of coastal British Columbia and adjacent areas. Rodrigues et al. (2018) conducted the ancient DNA studies mentioned in this chapter; the pre-colonial human population estimate for Barkley Sound (8,500 people) is as described in that study. The description of technologies used by ancient fishers is based on personal communications with archaeologist Iain McKechnie, from the University of Victoria. Yamanaka and Logan (2010) documented the history of rockfish overexploitation in British Columbia and the conservation strategies initiated in the early 2000s to promote recovery.

The study on the modern chemistry of ancient trilobites is by John et al. (2017). The 41% average value for biogenic biomarkers in the trilobite specimens is the mean of "protected" and "unprotected" samples reported in Table 1 of that study. Sally Walker is quoted from a non-technical summary of this research in *Anthropocene* magazine (Keim 2017).

For a non-technical introduction into the scientific basis for the Anthropocene, see Revkin (2016). Over the past century and through our use of chemical fertilizers, we have caused the biggest shift to the nitrogen cycle in the past two-and-half-billion years. During the same period and via our burning of fossil fuels, we have caused the fastest rate of global warming in the past 66 million years. These are among the many lines of evidence that define the end of the Holocene and the start of the Anthropocene (Waters et al. 2016). Yet the key requirement for the Anthropocene to be formally accepted by scientists as a new geologic epoch is that of sweeping changes to the Earth's sediments and glaciers that are clearly and globally distinguishable from those occurring during earlier periods. Those changes have occurred. As Waters et al. (2016) wrote in their technical synthesis, "Recent anthropogenic deposits, which are the products of mining, waste disposal (landfill), construction, and urbanization, contain the greatest expansion of new minerals since the Great Oxygenation Event at 2400 [million years ago] and are accompanied by many new forms of 'rock,' in the broad sense of geological materials with the potential for long term persistence."

Jennifer Jacquet is quoted from Jacquet (2013).

Chapter 2: Resisting Least Resistance

For research on hydrodynamic trails made by fish and the ability of pinnipeds to follow them, see Denhardt et al. (2001). For a summary of how stable isotopes have been used to trace salmon-derived nutrients in the forest, see Schindler et al. (2003). Miriam Rothschild is quoted by Hugh Iltis, who attributes the original quote to a Nova television interview (Iltis 1988).

For technical introductions to ecological resilience, see Holling (1973), Folke et al. (2004), and Desjardins et al. (2015). For a non-technical introduction, see Zolli (2013). Berkes et al. (1998, 2000) discuss how traditional Indigenous societies applied the concept of ecological resilience long before Western science became aware of it.

Works describing relationships between Indigenous and settler cultures include Treuer (2019), Talaga (2018), Kimmerer (2014), King (2012), Sellars (2013), Johnson (2016) and Brody (2001). For an example of research on extinctions caused by human hunters millennia ago, see Saltré (2016). Non-technical works describing species extinctions caused by ancient hunters or traditional societies include Diamond (2005) and Harari (2015).

Evidence for a 1°C rise in global temperature over the past century is summarized in a public communication by climatologist James Hansen and colleagues (2018).

As stated by Berkes et al. (2006), the term "roving bandit" was coined by "[t]he economist Mancur Olson [who] argued that local governance creates a vested interest in the maintenance of local resources, whereas the ability of mobile agents—roving bandits in Olson's terminology—to move on to other, unprotected resources severs local feedback and the incentive to build conserving institutions."

Works presenting evidence or examples for the historical and long-term capacity of many traditional cultures to learn to live sustainably include Berkes (2018), Kimmerer (2014), Turner (2005), Brody (2001), Mathews and Turner (2017), Berkes and Turner (2006) and Turner and Berkes (2006). Importantly, Berkes and Turner (2006) state, "Conservation does not come naturally; it has to be learned," and provide examples of how different societies have undertaken that learning, or failed to do so. See also Diamond (2005).

For a discussion on the regaining of political power by Indigenous people in Canada, see Saul (2014). For a parallel discussion focused on the United States, see Treuer (2019).

Dúqvàísḷa (William Housty) is quoted from his paper (written with collaborators) integrating Western science and traditional laws in the monitoring and conservation of grizzly bears (Housty et al. 2014). The statement on Indigenous people forming 5% of the current global population is based on 2018 statistics from the World Bank (World Bank 2018). Robin Wall Kimmerer is quoted from her book, *Braiding Sweetgrass* (Kimmerer 2014).

Chapter 3: Coalescing Knowledge

The opening paragraphs of this chapter, concluding that the loss or suppression of Indigenous cultures is narrowing our collective psyche, are informed by Berkes (2018), Kimmerer (2014), Davis (2009), Turner (2005) and Brody (1992, 2001). To learn about the history of residential schools and other forms of institutionalized racism against Indigenous peoples in Canada, see Talaga (2018) and reports from the National Centre for Truth and Reconciliation (Truth and Reconciliation 2019). For a firsthand account of a residential school survivor, see Sellars (2013). Quotes by Indigenous peoples interviewed by Lauren Eckert are as published in Eckert et al. (2018).

For a global synthesis on the integration of Indigenous traditional knowledge and Western science, see Berkes (2018). For writings on this topic from the positionality of an Indigenous scientist, see Kimmerer (2002, 2014). Some of the interpretations I present in this chapter were first published in Ban, Frid et al. (2018). Artelle et al. (2018) integrate the values and beliefs of place-based traditional cultures and the perspectives of Western scientists into an argument for "values-led management" of natural resources. For an introduction to marine microbial diversity and the ecological role of marine microbes, see Salazar and Sunagawa (2017).

The estimate of 102,000 to 210,000 Indigenous people inhabiting the Northwest Coast (coastal areas of Southeast Alaska, British Columbia, Washington and Oregon) when Europeans arrived is as reported by Campbell and Butler (2010), who cite primary research by forensic anthropologist D.H. Ubelaker. Works on the imbrication of archaeological evidence and oral narratives referred to in this chapter are by Gauvreau and McLaren (2016) and McLaren et al. (2015). The latter discusses evidence for 10,000 years of human occupation at four sites and at least 5,000 years at five other sites of British Columbia's Central Coast. Archaeological evidence for the portfolio of marine resources used by Indigenous peoples of the Northeast Pacific, and how different resources ranked in relative importance, is as reported by McKechnie and Moss (2016). Evidence for 7,000 years of use of salmon fisheries and storage technologies at the village of Namu is as reported by Cannon and Yang (2006). The research into stone wall traps for salmon is as reported by Heiltsuk archaeologist, Gitla (Elroy White) (White 2006); see also a video on his work (Tyee 2012). The example of Sugpiaq people shifting their harvesting behavior when electricity and freezers were first introduced to their village, and the accompanying quote, are as reported by Salomon et al. (2007). Campbell and Butler (2010) further discuss the archaeological evidence for the longevity of Indigenous salmon fisheries over the course of millennia and the role of social institutions in achieving sustainability and ecological resilience in these fisheries.

For a synthesis on serial depletion by industrial fisheries, see Pauly et al. (2002). For the history of rockfish over-exploitation in British Columbia and management measures to reverse downward trends, see Yamanaka and Logan (2010). Our studies documenting the declines in average size and age of yelloweye and quillback rockfish include interviews with traditional fishers (Eckert et al. 2017) and analyses of scientific survey data (McGreer and Frid 2017).

The *Kitasoo/Xai'xais Management Plan For Pacific Herring* (Kitasoo/Xai'xais First Nation 2019) is available online from the marine planning webpage of the Kitasoo/Xai'xais Resource Stewardship Office. The plan is updated annually, so different versions will appear in future years.

For a synthesis of stewardship, harvesting, and tending practices by coastal First Nations (including use of clam gardens, fish traps, and estuarine root gardens), see Mathews and Turner (2017). The meaning of *ksnmsta* is defined in the book *Alhqulh Ti Sputc*, produced in 2017 by the Nuxalk Nation to compile their knowledge on eulachon (*Sputc*); the book was created for internal use and is not publicly available. Sources describing the hereditary chief system include Trosper (2003) and Kitasoo/Xai'xais First Nation (2019). For an introduction to clam gardens, see Jackley et al. (2016) and Mathews and Turner (2017).

For experimental tests of how clam gardens affect biological productivity, see Groesbeck et al. (2014). The 3,500 year estimate for the age of clam gardens is as reported by Smith et al. (2019). Uses of dogfish mentioned in this chapter are as described by McKechnie and Moss (2016). For work on clam garden restoration and its cultural and ecological significance, see Augustine and Dearden (2014) and the website of the The Clam Garden Network (Clam Garden Network 2019). Thomas King is paraphrased from his book, *The Inconvenient Indian* (King 2012).

Chapter 4: Reawakening

For an introduction to the Wuikinuxv Nation, see the booklet produced in association with the University of British Columbia Museum of Anthropology (Wuikinuxv Nation 2011).

Throughout most of this book I refer to the smelt *Thaleichthys pacificus* by the common name of "eulachon," which is consistent with the scientific literature. As depicted by quotes that appear in this chapter, however, Coastal First Nations, generally refer to the same species as "ooligan."

I owe much of my understanding of eulachon to my colleague Megan Moody. Her graduate thesis in fisheries (Moody 2008), which is written from the positionality of a Nuxalkmc scientist and which I quote in this chapter, documents the collapse, ecology, conservation, and cultural significance of eulachon; it also explains the process of turning eulachon into grease, which is fundamental to the culture of many First Nations. Kuhnlein et al. (1982) also explain the nutritional qualities of eulachon grease and its production. For additional information on the shrimp trawl fishery see Fisheries and Oceans Canada (2019a).

Research by Robert Hannah and colleagues on the use of LED lights to reduce bycatch is published in Hannah et al. (2015), from which I quote. The Oregon Department of Fish and Wildlife is quoted from its website (Oregon Department of Fish and Wildlife 2019).

Chapter 5: The Exuberance of Herring

To learn more about the relationship between Pacific herring and the Kitasoo/Xai'xais and Heiltsuk peoples, see, respectively, Kitasoo/Xai'xais First Nation (2019) and Gauvreau et al. (2017). The narrative of Neasmuutk Haimas's (Charlie Mason) upbringing is quoted with permission from the Kitasoo/Xai'xais archive, which is created and managed by the Nation and is not accessible to the general public. For archaeological research into the longevity and stability of herring fisheries prior to colonization, see McKechnie et al. (2014).

For entry points into the ecological role and vulnerability to industrial exploitation of herring and other forage fishes, see Pikitch et al. (2012) and Essington et al. (2015). For research into the greater sustainability of harvesting herring eggs rather than adults, see Shelton et al. (2014). My summary of the history of herring exploitation in British Columbia is based on Ware (1985) and Fisheries and Oceans Canada (1994).

The research on herring spawning at deeper depths is currently published as the thesis of Markus Thompson (2017), yet more publications are soon to be available. For Heiltsuk observations on the declining number and shrinking geographic distribution of herring spawning sites, see Gerrard (2014). My statement that some Heiltsuk believe "juveniles may be making some poor choices when it comes to choosing their spawning locations" is based on interviews reported by Gauvreau et al. (2017).

For an introduction to The Blob and other sources of variability in ocean temperatures, see Chavez et al. (2017). My statement about sea surface temperatures in British Columbia's Central Coast being nearly three degrees warmer, on average, at the time of herring spawn (March 20-April 7) during 2015 and 2016 (mean 10.01°C) relative to the preceding 60 years (mean 7.19°C) is based on annual time-series data for McInnes Island (Fisheries and Oceans Canada 2019b). For a synthesis on the synergistic effects of climate and fisheries on marine ecosystems, see Cheung (2018).

Events pertaining to the standoff between the Heiltsuk Nation and Fisheries and Oceans Canada in 2015 are as described in the Heiltsuk Nation's website (see Heiltsuk Nation 2015a, 2015b). DFO's quoted announcement concerning the Central Coast stock assessment area was sent by Victoria Postlethwaite in an e-mail message to the author and other recipients on March 1, 2018.

Chapter 6: Sculpted by River and Story

Data on concentrations of atmospheric carbon dioxide (CO_2) since 1958 and on annual rates of CO_2 emissions are as recorded by the US National Oceanic and Atmospheric Administration (Global Greenhouse Gas Reference Network 2019). My statement, "At the time of this writing, we have reached 415 parts per million of CO_2..." reflects data for May 13–17 of 2019.

Solomon et al. (2009) describe CO2 levels in the mid-1700s, the long-term persistence of CO2 in the atmosphere, and warming lags. Tripati et al. (2009) describe CO2 levels and climate from 15–20 million years ago. Predictions on glacial melt and associated changes to water discharge in western Canada are as reported by Clarke et al. (2015).

To learn about the decline of Atnarko sockeye and the efforts by the Nuxalk

Nation to recover this salmon stock, see Connors and Atnarko Sockeye Recovery Planning Committee (2016). For research on the relationship between salmon declines and increased rates of bears killed to protect people, see Artelle et al. (2016). For original works on "the tyranny of small decisions," see Kahn (1966) and Odum (1982).

Nuxalkmc lawyer Andrea Hilland and the Nuxalk origin story are quoted from Hilland (2013). The origin story presented by Hilland was originally recorded by anthropologist Thomas McIlwraith. Importantly, Hilland adds the following footnote to McIlwraith's version:

> McIlwraith uses masculine nouns and pronouns to refer to both men and women. Although the use of "man" to describe humans is archaic, sexist, and ambiguous, I have quoted the authors directly with the proviso that, as a Nuxalk citizen, I know that Nuxalk women are included as first ancestors and as chiefs in Nuxalk society.

Robin Wall Kimmerer is quoted from Kimmerer (2014).

Chapter 7: Beautiful Protest

Our research on Dungeness crab described in this chapter is available in the scientific literature (Ban et al. 2017; Frid et al. 2016b), and summarized—with policy implications—by Ban, Frid et al. (2018). My discussion on the complementarity between qualitative signals used by traditional knowledge holders and quantitative measures used by scientists is informed by Berkes and Berkes (2009), Chapter 9 of Berkes (2018), and Brody (1992). Anna Badkhen is quoted from her book, *Fisherman's Blues* (Badkhen 2018). Nancy Turner is quoted from Turner (2003). Henrik Moller, Fikret Berkes, and colleagues are paraphrased from Moller et al. (2004).

Chapter 8: Echoes Across the Lake

For an introduction to the Tŝilhqot'in court case, see materials in the website of the Tŝilhqot'in National Government (Tŝilhqot'in National Government 2019). Marylin Baptiste's 2010 op-ed appeared in the *Globe and Mail* (Baptiste 2010).

I was not present at the pole-raising ceremony that took place at Teztan Biny during October of 2014. My writing of that event is based on conversations with Cecil Grinder and the documentary video created by the Tŝilhqot'in National Government with filmmaker Dale Devost (Tŝilhqot'in National Government 2015). Nuu-Chah-Nulth carver Tim Paul is quoted from that video.

For insights into First Nations, substance abuse and healing, written from the positionality of an Indigenous lawyer and writer, see Harold R. Johnson's *Firewater* (Johnson 2016).

Chapter 9: Ditching Our Climate-Wrecking Stories

Quotes by Bob Dudley and related data are from the BP Statistical Review of World Energy 2017 (BP 2017). For a 2016 film addressing the declining return on investment for fossil fuels, see David Lavallée's *To the Ends of the Earth* (Lavallée 2019). Petronas' withdrawal from their natural gas project proposal was reported by Paraskova (2017). For documentation of Rick Perry's swings in support for different kinds of energy, see Tollefson (2016), Hand (2017), Gstalter (2018), and Green (2018).

"Roadmap for Rapid Decarbonization" and "Three Years To Safeguard Our Climate" are available in the scientific literature (Rockström et al. (2017), Figueres et al. (2017)). Hans Joachim Schellnhuber is quoted from Carrington (2015).

News headlines on the extreme fires of 2017 and 2018 in British Columbia are quoted from articles by CBC (CBC News 2017) and Global News (Judd 2018). The British Columbia Wildland Fire Management Strategy is available from the BC Wildfire Service website (BC Wildfire Service 2010). The financial costs of forest fires up to 2015 in British Columbia are based on news articles by CBC (CBC News 2015) and Global News (Talmazan and Dunn 2015). The description of the 2018 wildfire season in British Columbia is quoted from BC Wildfire Service (2019).

The figure for Canadian subsidies to the fossil fuel industry estimated in 2016 was reported in The Guardian (Milman 2016); while the absolute value of those subsidies has declined, a 2018 study concluded that "Canada is still the largest provider of subsidies to oil and gas production in the G7 per unit of GDP" (Touchette and Gass (2018), Whitley et al. (2018)). The figures I provide on declines in global fossil fuel subsidies are as reported from the International Energy Agency World Energy Outlook Reports (IEA 2019).

Springmann et al. (2016) quantify the climate costs of different diets. Anderson (2012) provides insights into the climate costs of jet flying.

Chapter 10: At the Edge of Geological Epochs

Turner (2003) describes the ethnobotany of the seaweed *Pyropia* (formerly *Porphyra*); quotes by Hellen Clifton and Indigenous names for *Pyropia* are as presented in that work. For an example on how small variations in temperature can affect productivity of a seaweed (bull kelp, in this case), see Krumhansl et al. (2017). My statement about April sea surface temperatures at McInnes Island averaging almost three degrees warmer during 2015 and 2016 (mean 10.63°C) relative to the preceding 60 years (mean 7.87°C) is based on annual time-series data available from Fisheries and Oceans Canada (Fisheries and Oceans Canada 2019b). For an introduction to The Blob and other sources of variability in ocean temperatures, see Chavez et al. (2017).

The disappearance and return of eulachon observed by the Nuxalk Nation is as described in the book *Alhqulh Ti Sputc* (not publicly available). The disappearance and return of caribou observed by the Cree of northeastern Canada is as described in Chapter 5 of Berkes (2018). Modern extinction rates and use of the term "biological annihilation" are as reported by Ceballos et al. (2017). For references on ecological resilience, see notes for Chapter 2.

Our studies on recent and historical changes to the body sizes of rockfish are available in the scientific literature (Eckert et al. 2017; McGreer and Frid 2017). The notes for Chapter 1 provide references for rockfish life history characteristics, overexploitation, and longevity overfishing.

To learn about the effects of ocean warming on metabolism and growth rates of fish, see Cheung et al. (2012). John Terborgh is quoted from his seminal synthesis on the relationship between top predators and biodiversity (Terborgh 2015). For a synthesis on the ecological effects of predators in the ocean and the consequences of their loss, see Heithaus et al. (2008); additionally Rizzari et al. (2014) and Frid et al. (2012) used field experiments to test predictions on predator-prey behavioral interactions and the ecological consequences of losing large predators. To learn more about the ecological role of lingcod, see Beaudreau and Essington (2007).

Scientists of the Anthropocene Working Group of the International Commission on Stratigraphy are quoted from Waters et al. (2016). To learn about the effects of multiple, synergistic stressors on ecosystems, see Brook et al. (2008).

The ongoing process of creating a marine protected area network in the Northern Shelf Bioregion of British Columbia (which includes the Central Coast) is described at MPA Network (2019). For an example of the contributions of our research to that process, see Frid et al. (2018).

The concluding section of this chapter is based on a personal journey through the Edziza and Spectrum mountain ranges of northern British Columbia, which occurred while extreme forest fires burned nearby in the Alkali Lake and South Stikine River areas; the size of these merged fires is as reported by the BC Wildfire Service (2019). To learn about the historical use of obsidian by the Tahltan and other aspects of the archaeology and geological history of that region, see Fladmark (1984) and Reimer (2015).

Chapter 11: Transformation

Dunham et al. (2018) and Kahn et al. (2018) report research on sponge bioherms—deep water, biogenic habitats that play important roles in water filtering and carbon sequestration—in Howe Sound and vicinity. Kahn et al. used stable isotopes to trace the terrestrial and marine nutrient subsidies that support the bacteria consumed by cloud sponges. The Halkett bioherm described at the beginning of the chapter is difficult to access; I thank my friends Adam Taylor and Glenn Dennison, from the Marine Life Sanctuaries Society (MLSS), for inviting me to tag along with them for my most recent set of dives there. To learn more about MLSS's work documenting and protecting bioherms in Howe Sound, see Marine Life Sanctuaries Society (2019).

The correspondence from Mike Reid quoted in this chapter alludes to the work of forest ecologist Suzanne Simard, who has conducted an incredible body of work on how individual trees communicate with each other and share resources through mycorrhizal networks. To learn about her work, see University of British Columbia (2019).

For research into declining trends in global violence, see Pinker (2012).

Nancy Turner and Fikret Berkes are quoted from two of their collaborative papers (Berkes and Turner (2006); Turner and Berkes (2006). For a classic paper on optimal foraging, see Charnov (1976). For an example of central place foraging and the depletion halo in seabirds, see Elliott et al. (2009). For our work applying central pace foraging to sports fishers, see Frid et al. (2016a); other examples include Bellquist and Semmens (2016) and Haggarty et al. (2016). Houston (2011) presents a theoretical treatise of central place foraging applied to humans. For an example of optimal foraging theory applied to Indigenous hunters in the neotropics, see Alvard et al. (1995). The concept of "roving bandits" is as presented by Berkes et al. (2006). The archaeological research in Haidà Gwaii described in this chapter is by Szpak et al. (2013). To learn more about the equation that encapsulates all known physical laws in the universe, see Turok (2012).

References

Alvard, Michael, Janis B. Alcorn, Richard E. Bodmer, Raymond Hames, Kim Hill, Jean Hudson, R. Lee Lyman et al. 1995. "Intraspecific prey choice by Amazonian hunters [and comments and reply]." *Current Anthropology* 36, no. 5 (December): 789–818.

Anderson, Kevin. 2012. "The inconvenient truth of carbon offsets." *Nature* 484, no, 7392 (April): 7.

Artelle, Kyle A., Janet Stephenson, Corey Bragg, Jessie A. Housty, William G. Housty, Merata Kawharu, and Nancy J. Turner. 2018. "Values-led management: the guidance of place-based values in environmental relationships of the past, present, and future." *Ecology and Society* 23, no. 3: 35. doi.org/10.5751/ES-10357-230335.

Artelle, Kyle A., Sean C. Anderson, John D. Reynolds, Andrew B. Cooper, Paul C. Paquet, and Chris T. Darimont. 2016. "Ecology of conflict: marine food supply affects human-wildlife interactions on land." *Scientific Reports* 6 (May): 25936.

Augustine, Skye, and Philip Dearden. 2014. "Changing paradigms in marine and coastal conservation: A case study of clam gardens in the Southern Gulf Islands, Canada." *Canadian Geographer / Le Géographe Canadien* 58, no. 3 (February): 305–314.

Badkhen, Anna. 2018. *Fisherman's Blues: A West African Community at Sea.* New York, NY: Riverhead Books.

Ban, Natalie C., Lauren Eckert, Madeleine McGreer, and Alejandro Frid. 2017. Indigenous knowledge as data for modern fishery management: a case study of Dungeness crab in Pacific Canada. *Ecosystem Health and Sustainability* 3, no. 8: 1379887.

Ban, Natalie Corinna, Alejandro Frid, Mike Reid, Barry Edgar, Danielle Shaw, and Peter Siwallace. 2018. "Incorporate Indigenous perspectives for impactful research and effective management." *Nature Ecology & Evolution* 2, no. 11 (October): 1680–1683.

Baptiste, Marilyn. 2010. "Ottawa must not sell out Tŝilhqot'in." *Globe and Mail.* Updated April 28, 2018. theglobeandmail.com/opinion/ottawa-must-not-sell-out-tsilhqotin/article1241117/?arc404=true.

Barnett, Lewis A. K., Trevor A. Branch, R. Anthony Ranasinghe and Timothy E. Essington. 2017. "Old-growth fishes become scarce under fishing." *Current Biology* 27, no. 18 (September): 2843–2848.e2.

BC Wildfire Service. 2010. "Wildland fire management strategy." September 2010, accessed May 26, 2019. gov.bc.ca/assets/gov/public-safety-and-emergency-services/wildfire-status/governance/bcws_wildland_fire_mngmt_strategy.pdf.

BC Wildfire Service. 2019. "2018 wildfire season summary." Accessed May 8, 2019. gov.bc.ca/gov/content/safety/wildfire-status/about-bcws/wildfire-history/wildfire-season-summary.

Beamish, Richard J., Gordon A. McFarlane, and Ashleen J. Benson. 2006. "Longevity overfishing." *Progress in Oceanography* 68, no. 2 (February): 289–302.

Beaudreau, Anne H., and Thomas E. Essington. 2007. "Spatial, temporal, and ontogenetic patterns of predation on rockfishes by lingcod." *Transactions of the American Fisheries Society* 136, no. 5 (September): 1438–1452.

Bellquist, Lyall, and Brice X. Semmens. 2016. "Temporal and spatial dynamics of 'trophy'-sized demersal fishes off the California (USA) coast, 1966 to 2013." *Marine Ecology Progress Series* 547 (April): 1–18.

Berkeley, Steven A., Mark A. Hixon, Ralph J. Larson, and Milton S. Love. 2004. "Fisheries sustainability via protection of age structure and spatial distribution of fish populations." *Fisheries* 29, no. 8 (August): 23–32.

Berkes, Fikret. 2018. *Sacred Ecology*, 4th ed. New York: Routledge.

Berkes, Fikret, and Mina Kislalioglu Berkes. 2009. "Ecological complexity, fuzzy logic, and holism in indigenous knowledge." *Futures* 41, no. 1 (February): 6–12.

Berkes, Fikret, Johan Colding, and Carl Folke. 2000. "Rediscovery of traditional ecological knowledge as adaptive management." *Ecological Applications* 10, no. 5 (October): 1251–1262.

Berkes, F., T. P. Hughes, R. S. Steneck, J. A. Wilson, D. R. Bellwood, B. Crona, C. Folke et al. 2006. "Globalization, roving bandits, and marine resources." *Science* 311, no. 5767 (March): 1557–1558.

Berkes, Fikret, Mina Kislalioglu, Carl Folke, and Madhav Gadgil. 1998. "Exploring the basic ecological unit: Ecosystem-like concepts in traditional societies." *Ecosystems* 1: 409–415.

Berkes, Fikret, and Nancy J. Turner. 2006. "Knowledge, learning and the evolution of conservation practice for social-ecological system resilience." *Human Ecology* 34, no. 4 (October): 479–494.

BP. 2017. "BP statistical review of world energy." June 2017, accessed April 16,

2019. scribd.com/document/351869273/Bp-Statistical-Review-of-World-Energy-2017-Full-Report.

Brody, Hugh. 1992. *Maps and Dreams*. Vancouver: Douglas & McIntyre.

Brody, Hugh. 2001. *The Other Side of Eden: Hunters, Farmers and the Shaping of the World*. Vancouver: Douglas & McIntyre.

Brook, Barry W., Navjot S. Sodhi, and Corey J.A. Bradshaw. 2008. "Synergies among extinction drivers under global change." *Trends in Ecology and Evolution* 23, no. 8 (August): 453–460.

Campbell, Sarah K., and Virginia L. Butler. 2010. "Archaeological evidence for resilience of Pacific Northwest salmon populations and the socio-ecological system over the last ~7,500 years." *Ecology and Society* 15, no. 1: 17. ecologyandsociety.org/vol15/iss1/art17/.

Cannon, Aubrey, and Dongya Yang. 2006. "Early storage and sedentism on the Pacific Northwest Coast: Ancient DNA analysis of salmon remains from Namu, British Columbia." *American Antiquity* 71, no. 1 (January): 123–140.

Carrington, Damian. 2015. "Fossil fuel industry must 'implode' to avoid climate disaster, says top scientist." *The Guardian*, July 10, 2015. theguardian.com/environment/2015/jul/10/fossil-fuel-industry-must-implode-to-avoid-climate-disaster-says-top-scientist.

CBC News. 2015. "B.C. balances budget despite $380M bill for forest fires." Last updated September 15, 2015. cbc.ca/news/canada/british-columbia/b-c-balances-budget-despite-380m-bill-for-forest-fires-1.3229211.

CBC News. 2017. "More than 45,000 people displaced by B.C. wildfires." Last updated July 18, 2017. cbc.ca/news/canada/british-columbia/bc-wildfires-tuesday-1.4210370.

Ceballos, Gerardo, Paul R. Ehrlich, and Rodolfo Dirzo. 2017. "Biological annihilation via the ongoing sixth mass extinction signaled by vertebrate population losses and declines." *Proceedings of the National Academy of Sciences* 114, no. 30 (July): E6089-E6096.

Charnov, Eric L. 1976. "Optimal foraging, the marginal value theorem." *Theoretical Population Biology* 9, no. 2 (April): 129–136.

Chavez, Francisco, J. Timothy Pennington, Reiko Michisaki, Margaret Blum, Gabriela Chavez, Jules Friederich, Brent Jones et al. 2017. "Climate variability and change response of a coastal ocean ecosystem." *Oceanography (Washington D.C.)* 30, no. 4 (December): 128–145.

Cheung, William W.L. 2018. "The future of fishes and fisheries in the changing oceans." *Journal of Fish Biology* 92, no. 3 (March): 790–803.

Cheung, William W.L., Jorge L. Sarmiento, John Dunne, Thomas L. Frölicher, Vicky W.Y. Lam, M.L. Deng Palomares, Reg Watson et al. 2012.

"Shrinking of fishes exacerbates impacts of global ocean changes on marine ecosystems." *Nature Climate Change* 3: 254–258.

Clarke, Garry K. C., Alexander H. Jarosch, Faron S. Anslow, Valentina Radić, and Brian Menounos. 2015. "Projected deglaciation of western Canada in the twenty-first century." *Nature Geoscience* 8: 372–377.

Clam Garden Network. "Eco-cultural restoration and the Gulf Islands National Park Reserve." Accessed April 4, 2019. clamgarden.com/research-2/restoration-in-gulf-islands/.

Connors, B. M., and Atnarko Sockeye Recovery Planning Committee. 2016. "Atnarko Sockeye Recovery Plan." July 26, 2016. ccira.ca/wp-content/uploads/2018/07/AtnarkoSockeyRecoveryPlan-FullSizeRender-45.pdf.

Davis, Wade. 2009. *The Wayfinders: Why Ancient Wisdom Matters in the Modern World.* Toronto: House of Anansi Press.

Dehnhardt, Guido, Björn Mauck, Wolf Hanke, and Horst Bleckmann. 2001. "Hydrodynamic trail-following in harbor seals (*Phoca vitulina*)." *Science* 293, no. 5527 (July): 102–104.

Desjardins, Eric, Gillian Barker, Zoë Lindo, Catherine Dieleman, and Antoine C. Dussault. 2015. "Promoting resilience." *Quarterly Review of Biology* 90, no. 2 (June): 147–165.

Diamond, Jared. 2005. *Collapse: How Societies Choose to Fail or Succeed.* New York: Penguin.

Dick, Edward J., Sabrina G. Beyer, Marc Mangel, and Stephen Ralston. 2017. "A meta-analysis of fecundity in rockfishes (genus *Sebastes*)." *Fisheries Research* 187 (March): 73–85.

Dunham, A., S. K. Archer, S. C. Davies, L. A. Burke, J. Mossman, J., J. R. Pegg, and E. Archer. 2018. "Assessing condition and ecological role of deep-water biogenic habitats: Glass sponge reefs in the Salish Sea." *Marine Environmental Research* 141: 88–99.

Eckert, Lauren E., Natalie C. Ban, Alejandro Frid, and Madeleine McGreer. (2017). Diving back in time: Extending historical baselines for yelloweye rockfish with Indigenous knowledge. *Aquatic Conservation Marine and Freshwater Ecosystems* 28, no. 1 (February): 158–166.

Eckert, Lauren E., Natalie C. Ban, NSnxakila-Clyde Tallio, and Nancy Turner. 2018. "Linking marine conservation and Indigenous cultural revitalization: First Nations free themselves from externally imposed social-ecological traps." *Ecology and Society* 23, no. 4: 23.

Elliott, Kyle, Kerry Woo, Anthony J. Gaston, Silvano Benvenuti, Luigi Dall'-Antonia, and Gail Davoren. 2009. "Central-place foraging in an Arctic seabird provides evidence for Storer-Ashmole's halo." *Auk* 126, no. 3 (July): 613–625.

Essington, Timothy E., Pamela E. Moriarty, Halley E. Froehlich, Emma E. Hodgson, Laura E. Koehn, Kiva L. Oken, Margaret C. Sipple et al. 2015. "Fishing amplifies forage fish population collapses." *Proceedings of the National Academy of Sciences* 112, No. 21 (May): 6648–6652.

Figueres, Christiana, Hans Joachim Schellnhuber, Gail Whiteman, Johan Rockström, Anthony Hobley, and Stefan Rahmstorf. 2017. "Three years to safeguard our climate." *Nature* 546, no. 7660 (June): 593–595.

Fisheries and Oceans Canada. 1994. "Herring spawn and catch records of British Columbia." Archived content, accessed May 19, 2019. pac.dfo-mpo.gc.ca/science/species-especes/pelagic-pelagique/herring-hareng/hertags/pages/indicator-eng.htm.

Fisheries and Oceans Canada. 2019a. "Shrimp trawl, Pacific Region 2019/2020 integrated fisheries management plan summary." Accessed April 10, 2019. pac.dfo-mpo.gc.ca/fm-gp/mplans/shrimp-trawl-crevette-chalut-ifmp-pgip-sm-eng.html

Fisheries and Oceans Canada. 2019b. "British Columbia lightstation sea-surface temperature and salinity data (Pacific), 1914–present." Accessed April 10, 2019. open.canada.ca/data/en/dataset/719955f2-bf8e-44f7-bc26-6bd623e82884.

Fladmark, Knut R. 1984. "Mountain of glass: Archaeology of the Mount Edziza obsidian source, British Columbia, Canada." *World Archaeology* 16, no. 2 (October): 139–156.

Folke, Carl, Steve Carpenter, Brian Walker, Martin Scheffer, Thomas Elmqvist, Lance Gunderson, and C. S. Holling . 2004. "Regime shifts, resilience, and biodiversity in ecosystem management." *Annual Review of Ecology, Evolution, and Systematics* 35 (December): 557–581.

Frid, Alejandro, Jeff Marliave, and Michael R. Heithaus. 2012. "Inter-specific variation in life history relates to antipredator decisions by marine mesopredators on temperate reefs." *PLoS One* 7, no. 6 (June): e40083.

Frid, Alejandro, Madeleine McGreer, Katie SP Gale, Emily Rubidge, Tristan Blaine, Mike Reid, Angeleen Olson et al. 2018. "The area–heterogeneity tradeoff applied to spatial protection of rockfish (*Sebastes* spp.) species richness." *Conservation Lettters* 2018: e12589.

Frid, Alejandro, Madeleine McGreer, Dana Haggarty, Julie Beaumont, and Edward J. Gregr. 2016a. "Rockfish size and age: The crossroads of spatial protection, central place fisheries and indigenous rights." *Global Ecology and Conservation* 8 (October): 170–182.

Frid, Alejandro, Madeleine McGreer, and Angela Stevenson. 2016b. "Rapid recovery of Dungeness crab within spatial fishery closures declared under indigenous law in British Columbia." *Global Ecology and Conservation* 6 (April): 48–57.

Gauvreau, Alisha M., and Duncan McLaren. 2016. "Stratigraphy and storytelling: Imbricating Indigenous oral narratives and archaeology on the Northwest Coast of North America." *Hunter Gatherer Research* 2, no. 3: 303–325.

Gauvreau, Alisha M., Dana Lepofsky, Murray Rutherford, and Mike Reid. 2017. "'Everything revolves around the herring:' the Heiltsuk-herring relationship through time." *Ecology and Society* 22, no. 2: 10.

Gerrard, Aniece Linée. 2014. "Understanding the past to inform future conservation policy: Mapping traditional ecological knowledge of Pacific herring spawning areas through time." M.R.M. (Planning) project, Simon Fraser University. summit.sfu.ca/item/14526.

Global Greenouse Gas Reference Network. "Trends in atmospheric carbon dioxide." NOAA Earth System Research Laboratory, Global Monitoring Division. Accessed May 18, 2019. esrl.noaa.gov/gmd/ccgg/trends/data.html.

Green, Miranda. 2018. "Rick Perry: US 'blessed' to provide fossil fuels to the world." *The Hill.* January 24, 2018. thehill.com/policy/energy-environment/370585-rick-perry-us-blessed-to-provide-fossil-fuels-to-the-world.

Groesbeck, Amy S., Kirsten Rowell, Dana Lepofsky, and Anne K. Salomon. 2014. "Ancient clam gardens increased shellfish production: Adaptive strategies from the past can inform food security today." *PLoS One* 9, no. 3 (March 11): e91235.

Gstalter, Morgan. 2018. "Perry calls global moves to shift from fossil fuels 'immoral.'" *The Hill.* March 18, 2018. thehill.com/policy/energy-environment/377420-perry-calls-global-shift-away-from-fossil-fuels-immortal.

Haggarty, Dana R., Steven J.D. Martell, and Jonathan B. Shurin. 2016. "Lack of recreational fishing compliance may compromise effectiveness of rockfish conservation areas in British Columbia." *Canadian Journal of Fisheries and Aquatic Sciences* 73, no. 10 (April): 1587–1598.

Hand, Mark. "Energy secretary targets wind, solar after overseeing renewables explosion as Texas governor." *Think Progress.* April 17, 2017. thinkprogress.org/energy-secretary-throws-bone-to-coal-nukes-with-review-of-wind-solar-subsidies-be58bc5af9f1.

Hannah, Robert W., Mark J.M. Lomeli, and Stephen A. Jones. 2015. "Tests of artificial light for bycatch reduction in an ocean shrimp (*Pandalus jordani*) trawl: Strong but opposite effects at the footrope and near the bycatch reduction device." *Fisheries Research* 170 (October): 60–67.

Hansen, James, Makiko Sato, Reto Ruedy, Gavin A. Schmidt, Ken Lob, and Avi Persin. "Global temperature in 2017." 18 January 2018. columbia.edu/~jehi/mailings/2018/20180118_Temperature2017.pdf.

Harari, Yuval Noah. 2015. *Sapiens: A Brief History of Humankind.* New York: Harper Collins.

Heiltsuk Nation. 2015a. "Heiltsuk Nation blockades DFO office on BC Central Coast: Nation prepares for direct action on the water as herring talks with DFO fail." March 30, 2015. heiltsuknation.ca/heiltsuk-nation-blockades-dfo-office-on-bc-central-coast/.

Heiltsuk Nation. 2015b. "Departure of empty gillnet boats from Central Coast validates Heiltsuk concerns over lack of herring: First Nation vacates DFO office as tensions ease." April 5, 2015. heiltsuknation.ca/departure-of-empty-gillnet-boats-from-central-coast-validates-heiltsuk-concerns-over-lack-of-herring/

Heithaus, Michael R., Alejandro Frid, Aaron J. Wirsing, and Boris Worm. 2008. "Predicting ecological consequences of marine top predator declines." *Trends in Ecology and Evolution* 23, no. 4 (April): 202–210.

Hilland, Andrea 2013. *Extinguishment By Extirpation: The Nuxalk Eulachon Crisis.* LL.M. thesis, University of British Columbia. open.library.ubc.ca/cIRcle/collections/ubctheses/24/items/1.0074234.

Hixon, Mark A., Darren W. Johnson, and Susan M. Sogard. 2014. "BOFFFFs: on the importance of conserving old-growth age structure in fishery populations." *ICES Journal of Marine Science* 71, no. 8 (October): 2171–2185.

Holling, C. S. 1973. "Resilience and stability of ecological systems." *Annual Review of Ecology and. Systematics* 4:1–23.

Houston, Alasdair .I. 2011. "Central-place foraging by humans: transport and processing." *Behavioral Ecology and Sociobiology* 65, no. 3 (March): 525–535.

Housty, William G., Anna Noson, Gerald W. Scoville, John Boulanger, Richard M. Jeo, Chris Darimont, and Christopher Filardi. 2014. "Grizzly bear monitoring by the Heiltsuk people as a crucible for First Nation conservation practice." *Ecology and Society* 19, no. 2 (June).

IEA. 2019. "World energy outlook reports." iea.org/weo/.

Iltis, Hugh H. 1988. "Serendipity in the exploration of biodiversity: What good are weedy tomatoes?" In *Biodiversity*, edited by E. O. Wilson, 98–105. Washington DC: National Academies Press. iltis.botany.wisc.edu/Serendipity%20in%20the%20exploration%20of%20biodiversity.html.

Jackley, Julia, Lindsay Gardner, Audrey Djunaedi, and Anne K. Salomon. 2016. "Ancient clam gardens, traditional management portfolios, and the resilience of coupled human-ocean systems." *Ecology and Society* 21, no. 4 (December): 20.

Jacquet, Jennifer. 2013. "The anthropocebo effect" *Conservation Biology* 27, no. 5 (October): 898–899.

John, Douglas L., Patricia M. Medeiros, Lydia Babcock-Adams and Sally E. Walker. 2017. "Cambrian trilobites as archives for Anthropocene biomarkers and other chemical compounds." *Anthropocene* 17 (March): 99–106.

Johnson, Harold R. 2016. *Firewater: How Alcohol Is Killing My People (and Yours)*. Regina SK: University of Regina Press.

Judd, Amy. 2018. "B.C. wildfire update Monday: More fires now burning than during record 2017 season." Global News. Last updated August 13, 2018. globalnews.ca/news/4384377/b-c-wildfire-update-monday/.

Kahn, Alfred E. 1966. "The tyranny of small decisions: market failures, imperfections, and the limits of economics." *Kyklos* 19, no. 1 (February): 23–47.

Kahn, Amanda S., Jackson W. F. Chu, and Sally P. Leys. 2018. "Trophic ecology of glass sponge reefs in the Strait of Georgia, British Columbia." *Scientific Reports* 8: 756.

Keim, Brandon. 2017. "The disturbingly modern chemistry of 500 million-year-old fossils." *Anthropocene* Magazine, Daily Science (April 19). anthropocenemagazine.org/2017/04/anthropocene-trilobites/

Kimmerer, Robin Wall. 2002. "Weaving traditional ecological knowledge into biological education: A call to action." *BioScience* 52, no 5 (May): 432–438.

Kimmerer, Robin Wall. 2014. *Braiding Sweetgrass: Indigenous Wisdom, Scientific Knowledge and the Teachings of Plants*. Minneapolis MN: Milkweed Editions.

King, Thomas. 2012. *The Inconvenient Indian: A Curious Account of Native People in North America*. Toronto: Doubleday Canada.

Kitasoo/Xai'xais First Nation. 2019. "Kitasoo/Xai'xais management plan for pacific herring." Klemtu, British Columbia, Canada. Download latest version at: klemtu.com/stewardship/marine-use-planning-management/.

Krumhansl, Kira A., Jordanna N. Bergman, and Anne K. Salomon. 2017. "Assessing the ecosystem-level consequences of a small-scale artisanal kelp fishery within the context of climate-change." *Ecological Applications* 27 (December): 799–813.

Kuhnlein, Harriet V., Alvin C. Chan, J. Neville Thompson, and Shuryo Nakai. 1982. "Ooligan grease: A nutritious fat used by Native people of Coastal British Columbia." *Journal of Ethnobiology* 2, no. 2 (December): 154–161.

Lavallée, David. 2019. *To the Ends of the Earth.* Accessed April 16, 2019. endsofearthfilm.com/.

Lin, J. D., M. A. Lemay, and L. W. Parfrey. 2018. Diverse bacteria utilize alginate within the microbiome of the giant kelp *Macrocystis pyrifera. Front. Microbiol.* 9:1914. doi: 10.3389/fmicb.2018.01914.

Love, Milton S., Mary Yoklavich, and Lyman Thorsteinson. 2002. *The Rockfishes of the Northeast Pacific.* Berkeley CA: University of California Press.

Mackie, Quentin, Daryl Fedje, Duncan McLaren, Nicole Smith, and Iain McKechnie. 2011. "Early environments and archaeology of Coastal British Columbia." In: *Trekking the Shore: Changing Coastlines and the Antiquity of Coastal Settlement*, ed. Nuno F. Bicho, Janathan A. Haws, and Loren G. Davis, 51–103. New York: Springer.

MacLaurin, Anne. 2017. "New find reaffirms Heiltsuk First Nation oral history—14,000 years ago," *UVic News* (April 11). uvic.ca/news/topics/2017+heiltsuk-alishagauvreau-ancient-village+ring.

Marine Life Sanctuaries Society. 2019. "Sponge research." Accessed April 19. 2019. mlssbc.com/programs/sponge-research/.

Mathews, Darcy L., and Nancy J. Turner. 2017. "Chapter 9—Ocean cultures: Northwest Coast ecosystems and Indigenous management systems." In *Conservation for the Anthropocene Ocean: Interdisciplinary Science in Support of Nature and People*, ed. Phillip S. Levin and Melissa R. Poe, 169–206. Amsterdam NL: Academic Press.

McGreer, Madeleine, and Alejandro Frid. 2017. "Declining size and age of rockfishes (*Sebastes* spp.) inherent to Indigenous cultures of Pacific Canada." *Ocean & Coastal Management* 145 (August): 14–20.

McKechnie, Iain, Dana Lepofsky, Madonna L. Moss, Virginia L. Butler, Trevor J. Orchard, Gary Coupland, Frederick Foster, et al. 2014. "Archaeological data provide alternative hypotheses on Pacific herring (*Clupea pallasii*) distribution, abundance, and variability." *Proceedings of the National Academy of Sciences* 111, No. 9 (February): E807-E816.

McKechnie, Iain, and Madonna L. Moss. 2016. "Meta-analysis in zooarchaeology expands perspectives on Indigenous fisheries of the Northwest Coast of North America." *Journal of Archaeological Science: Reports* 8 (August): 470–485.

McLaren, Duncan, Farid Rahemtulla, Gitla (Elroy White), and Daryl Fedje. 2015. "Prerogatives, sea level, and the strength of persistent places: Archaeological evidence for long-term occupation of the Central Coast of British Columbia." *B.C. Studies* 187 (Autumn): 155–191.

Milman, Oliver. 2016. "Canada gives $3.3bn subsidies to fossil fuel producers despite climate pledge." *The Guardian*, November 15, 2016. theguardian.com/world/2016/nov/15/climate-change-canada-fossil-fuel-subsidies-carbon-trudeau.

Moller, Henrik, Fikret Berkes, Philip O'Brian Lyver, and Mina Kislalioglu. 2004. "Combining science and traditional ecological knowledge: monitoring populations for co-management." *Ecology and Society* 9, no. 3 (November): 2. [online].

Moody, Megan Felicity. 2008. *Eulachon past and present.* M.Sc. thesis, University of British Columbia. open.library.ubc.ca/cIRcle/collections/ubctheses/24/items/1.0070785.

MPA Network. 2019. Accessed April 18, 2019. mpanetwork.ca/bcnorthernshelf/.

Odum, William E. 1982. "Environmental degradation and the tyranny of small decisions." *BioScience* 32, no. 9 (October): 728–729.

Oregon Department of Fish and Wildlife. "Commercial pink shrimp fishery—LED fishing light." Updated March 24, 2019. dfw.state.or.us/mrp/shellfish/commercial/shrimp/LEDs.asp.

Paraskova, Tsvetana. 2017. "Petronas bows out of $29B LNG project in Canada." Oilprice.com. July 26, 2017. oilprice.com/Latest-Energy-News/World-News/Petronas-Bows-Out-Of-29B-LNG-Project-In-Canada.html.

Pauly, Daniel, Villy Christensen, Sylvie Guénette, Tony J. Pitcher, Rashid Sumaila, Carl J. Walters, R. Watson et al. 2002. "Towards sustainability in world fisheries." *Nature* 418 (August), 689–695.

Pikitch, Ellen K., Pee D. Boersma, I. L. Boyd, David O. Conover, Philippe M. Cury, P., Timothy E. Essington, and Selina S. Heppell. 2012. *Little Fish, Big Impact: Managing a Crucial Link in Ocean Food Webs.* Washington DC: Lenfest Ocean Program.

Pinker, Steven. 2012. *The Better Angels Of Our Nature: Why Violence Has Declined.* New York: Penguin Publishing Group.

Reimer, Rudy. 2015. "Reassessing the role of Mount Edziza obsidian in northwestern North America." *Journal of Archaeological Science: Reports* 2 (June): 418–426.

Revkin, Andrew W. 2016. "An Anthropocene journey." *Anthropocene* Magazine (October). anthropocenemagazine.org/anthropocenejourney/.

Rizzari, Justin R., Ashley J. Frisch, Andrew S. Hoey, and Mark I. McCormick. 2014. "Not worth the risk: apex predators suppress herbivory on coral reefs." *Oikos* 123, no. 7 (July): 829–836.

Rockström, Johan, Owen Gaffney, Joeri Rogelj,Malte Meinshausen, Nebojsa Nakicenovic, and Hans Joachim Schellnhuber. 2017. "A roadmap for rapid decarbonization." *Science* 355, no 6331 (March): 1269–1271.

Rodrigues, Antonia, Iain McKechnie. and Dongya Yang. 2018. "Ancient DNA analysis of Indigenous rockfish use on the Pacific Coast: Implications for marine conservation areas and fisheries management." *PLoS One* 13, no. 2 (February): e0192716.

Salazar, Guillem, and Shinichi Sunagawa. 2017. "Marine microbial diversity." *Current Biology* 27, no. 11 (June): R489–R494.

Salomon, Anne K., Nick M. Tanape, and Henry P. Huntington. 2007. "Serial depletion of marine invertebrates leads to the decline of a strongly interacting grazer." *Ecological Applications* 17, no. 6 (October): 1752–1770.

Saltré, Frédérik, Marta Rodríguez-Rey, Barry W. Brook, Christopher N. Johnson, Chris S.M. Turney, John Alroy, Alan Cooper et al. 2016. "Climate change not to blame for late Quaternary megafauna extinctions in Australia." *Nature Communications* 7 (January): 10511.

Saul, John Ralston. 2014. *The Comeback: How Aboriginals Are Reclaiming Power And Influence*. Toronto ON: Penguin Random House Canada.

Schindler, Daniel E., Mark D. Scheuerell, Jonathan W. Moore, Scott M. Gende, Tessa B. Francis, and & Wendy L. Palen. 2003. "Pacific salmon and the ecology of coastal ecosystems." *Frontiers in the Ecolology and Environment* 1, no. 1 (February): 31–37.

Sellars, Bev. 2013. *They Called Me Number One: Secrets and Survival at an Indian Residential School*. Vancouver: Talonbooks.

Shelton, Andrew Olaf, Jameal F. Samhouri, Adrian C. Stier, and Philip S. Levin. 2014. "Assessing trade-offs to inform ecosystem-based fisheries management of forage fish." *Scientific Reports* 4: 7110.

Smith, Nicole F., Dana Lepofsky, Ginevra Toniello, Keith Holmes, Louie Wilson, L., Christina M. Neudorf, and Christine Roberts. 2019. "3500 years of shellfish mariculture on the Northwest Coast of North America." *PLoS ONE* 14, no. 2: e0211194. doi.org/10.1371/journal.pone.0211194.

Solomon, Susan, Gian-Kasper Plattner, Reto Knutti, and Pierre Friedlingstein. 2009. "Irreversible climate change due to carbon dioxide emissions." *Proceedings of the National Academy of Sciences* 106, no. 6 (February): 1704–1709.

Springmann, Marco, H. Charles J. Godfray, Mike Rayner, and Peter Scarborough. 2016. "Analysis and valuation of the health and climate change cobenefits of dietary change." *Proceedings of the National Academy of Sciences* 113, no. 15 (April 12): 4146–4151.

Szpak, Paul, Trevor J. Orchard, Anne K. Salomon, and Darren Gröcke. 2013. "Regional ecological variability and impact of the maritime fur trade on nearshore ecosystems in southern Haida Gwaii (British Columbia, Canada): evidence from stable isotope analysis of rockfish (*Sebastes* spp.) bone collagen." *Archaeological and Anthropological Science* 5, no. 2 (June): 159–182.

Talaga, Tanya. 2018. *All Our Relations: Finding the Path Forward.* Toronto: House of Anansi Press.

Talmazan, Yuliya, and Tavis Dunn. 2015. "INTERACTIVE: The cost of B.C. wildfires over the last decade." Global News. Last updated July 10, 2015. globalnews.ca/news/2101720/interactive-the-cost-of-b-c-wildfires-over-the-last-decade/.

Terborgh, John W. 2015. "Toward a trophic theory of species diversity." *Proceedings of the National Academy of Sciences* 112, no. 37 (September): 11415–11422.

Thompson, Markus D. 2017. "Causes and consequences of Pacific herring (*Clupea pallasii*) deep spawning behavior." M. R. M. research project, Simon Fraser University. summit.sfu.ca/item/17619.

Tollefson, Jeff. 2016. "Trump's pick for energy secretary once sought to eliminate DOE." *Nature*. Last updated December 14, 2016. nature.com/news/trump-s-pick-for-energy-secretary-once-sought-to-eliminate-doe-1.21160.

Touchette, Yanick, and Philip Gass. 2018. "Public cash for oil and gas: Mapping federal fiscal support for fossil fuels." International Institute for Sustainable Development. September 2018. iisd.org/sites/default/files/publications/public-cash-oil-gas-en.pdf.

Tripati, Aradhna K., Christopher D. Roberts, and Robert A. Eagle. 2009. "Coupling of CO_2 and ice sheet stability over major climate transitions of the last 20 million years." *Science* 326, no 5958: 1394–1397.

Trosper, Ronald L. 2003. "Resilience in pre-contact Pacific Northwest social ecological systems." *Ecology and Society* 7, no. 3: 6. consecol.org/vol7/iss3/art6/.

Treuer, David. 2019. *The Heartbeat of Wounded Knee: Native America From 1890 To The Present.* New York, NY: Riverhead Books.

Truth and Reconciliation Commission Reports. 2019. University of Manitoba. Accessed April 4, 2019. nctr.ca/reports2.php.

Tŝilhqot'in National Government. 2015. *Celebration for the Protection of Teztan Biny.*" Published May 22, 2015. youtu.be/CuTqIh-Ioj4.

Tŝilhqot'in National Government. 2019. "Tŝilhqot'in rights & title." Accessed April 15, 2019. tsilhqotin.ca/Tsilhqotin-Rights-Title.

Turner, Nancy J. 2003. "The ethnobotany of edible seaweed (*Porphyra abbottae* and related species; Rhodophyta: Bangiales) and its use by First

Nations on the Pacific Coast of Canada." *Canadian Journal of Botany* 81, no. 4: 283–293.

Turner, Nancy J. 2005. *The Earth's Blanket: Traditional Teachings for Sustainable Living.* Vancouver: Douglas & McIntyre.

Turner, Nancy J., and Fikret Berkes. 2006. "Coming to understanding: Developing conservation through incremental learning in the Pacific Northwest." *Human Ecology* 34, no. 4 (August): 495–513.

Turok, Neil. 2012. *The Universe Within: From Quantum to Cosmos.* Toronto ON: House of Anansi.

Tyee. "Stone fish traps explained." October 30, 2012. thetyee.ca/News/2012/10/30/BC-Central-Coast-Fish-Traps/.

University of British Columbia. 2019. Faculty of Forestry. Profiles. "Suzanne W. Simard." Accessed April 19, 2019. profiles.forestry.ubc.ca/person/suzanne-simard/.

Ware, D.M. 1985. "Life history characteristics, reproductive value, and resilience of Pacific herring (*Clupea harengus pallasi*)." *Canadian Journal of Fisheries and Aquatic Sciences* 42: s127–s139.

Waters, Colin N., Jan Zalasiewicz, Colin Summerhayes, Anthony D. Barnosky, Clément Poirier, Agnieszka Gałuszka, Alejandro Cearreta et al. 2016. "The Anthropocene is functionally and stratigraphically distinct from the Holocene." *Science* 351, no. 6269 (January): aad2622-1 to aad2622-10.

White (Xanius), Elroy. 2006. *Heiltsuk Stone Fish Traps: Products of My Ancestors' Labour.* M.A. diss, Simon Fraser University, summit.sfu.ca/item/4240.

Whitley, Shelagh, Han Chen, Alex Doukas, Ipek Gençsü, Ivetta Gerasimchuk, Yanick Touchette, and Leah Worrall. 2018. "G7 fossil fuel subsidy scorecard: Tracking the phase-out of fiscal support and public finance for oil, gas and coal," ODI. June 2018. odi.org/sites/odi.org.uk/files/resource-documents/12222.pdf.

World Bank. 2018. "Indigenous peoples." Last updated September 24, 2018. worldbank.org/en/topic/indigenouspeoples.

Wuikinuxv Nation. 2011. *We Are The Wuikinuxv Nation. UBC Museum Anthropol. Pacific Northwest Sourceb. Ser.*, 1–41. moa.ubc.ca/wp-content/uploads/2014/08/Sourcebooks-Wuikinuxv.pdf.

Yamanaka, Kae Lynne and Gary Logan. 2010. "Developing British Columbia's inshore rockfish conservation strategy." *Marine and Coastal Fisheries: Dynamics, Management, and Ecosystem System Science* 2 (January): 28–46.

Zolli, Andrew. 2013. "Good-bye sustainability, hello resilience." *Conservation* Magazine, University of Washington, (March 8). conservationmagazine.org/2013/03/good-bye-sustainability-hello-resilience/

Index

Note: cp indicates color section

A

B

C

About the Author

Alejandro Frid, PhD, has for over three decades inhabited the worlds of science, Indigenous cultures, and environmental activism. An ecologist for First Nations of British Columbia's Central Coast, and adjunct assistant professor in the School of Environmental Studies at the University of Victoria, Frid works collaboratively with First Nations on the integration of traditional knowledge and Western science to advance conservation and revitalize Indigenous control of their resources. His research experience has spanned conflicts between industrial development and terrestrial wildlife, the plight of endangered species, and the effects of overfishing on marine predators. Author of *A World for My Daughter*, he lives on Bowen Island, British Columbia.

ABOUT NEW SOCIETY PUBLISHERS

New Society Publishers is an activist, solutions-oriented publisher focused on publishing books for a world of change. Our books offer tips, tools, and insights from leading experts in sustainable building, homesteading, climate change, environment, conscientious commerce, renewable energy, and more—positive solutions for troubled times.

DON'T EAT THIS BOOK (but you could)

We're proud to hold to the highest environmental and social standards of any publisher in North America. This is why some of our books might cost a little more. We think it's worth it!

- We print all our books in North America, never overseas
- All our books are printed on **100% post-consumer recycled paper**, processed chlorine-free, with low-VOC vegetable-based inks (since 2002)
- Our corporate structure is an innovative employee shareholder agreement, so we're one-third employee-owned (since 2015)
- We're carbon-neutral (since 2006)
- We're certified as a B Corporation (since 2016)

At New Society Publishers, we care deeply about *what* we publish—but also about *how* we do business.

Download our catalog at https://newsociety.com/Our-Catalog or for a printed copy please email info@newsocietypub.com or call 1-800-567-6772 ext 111.

New Society Publishers

ENVIRONMENTAL BENEFITS STATEMENT

By using 100% post-consumer recycled paper vs virgin paper stock, New Society Publishers saves the following resources:[1] (per every 5,000 copies printed)

22	Trees
1,976	Pounds of Solid Waste
2,174	Gallons of Water
2,836	Kilowatt Hours of Electricity
3,592	Pounds of Greenhouse Gases
15	Pounds of HAPs, VOCs, and AOX Combined
5	Cubic Yards of Landfill Space

[1] Environmental benefits are calculated based on research done by the Environmental Defense Fund and other members of the Paper Task Force who study the environmental impacts of the paper industry.